Plug In!

AF585169

Plug In!

The Electrification Handbook

From the bestselling
author of **The Big Switch**

Saul Griffith
with Laura Fraser

Published by Black Inc.,
an imprint of Schwartz Books Pty Ltd
Wurundjeri Country
22–24 Northumberland Street
Collingwood VIC 3066, Australia
enquiries@blackincbooks.com
www.blackincbooks.com

Copyright © Saul Griffith 2025.
Saul Griffith and Laura Fraser assert their right to be known as the authors of this work.

ALL RIGHTS RESERVED.
No part of this publication may be reproduced, stored in a retrieval system, or transmitted in any form by any means electronic, mechanical, photocopying, recording or otherwise without the prior consent of the publishers.

9781760645151 (paperback)
9781743823927 (ebook)

A catalogue record for this book is available from the National Library of Australia

Book design and typesetting by Beau Lowenstern
Cover photograph by Clayton Boyd

Printed in Australia by McPherson's Printing Group.

Contents

Part 4
Frequently Asked Questions

200 KW

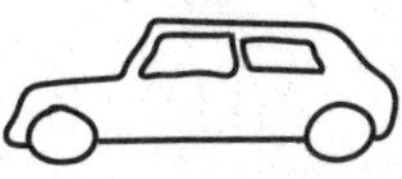

200,000 WATTS – TYPICAL FULL POWER OF AN ELECTRIC VEHICLE

100 W

100 WATTS – THE SUSTAINABLE EXTRA WORK OUTPUT OF A LABOURER

10 W

10 WATTS – POWER TO OPERATE A SINGLE LED LIGHTBULB

100 GW

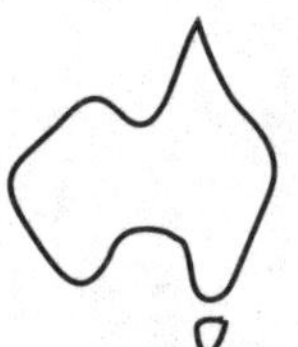

100 BILLION WATTS ARE REQUIRED TO KEEP AUSTRALIA HUMMING

1 KW

1000 WATTS ARE REQUIRED TO BURN YOUR TOAST

5 MW

A TYPICAL WIND TURBINE GENERATES 5 MILLION WATTS

100 W

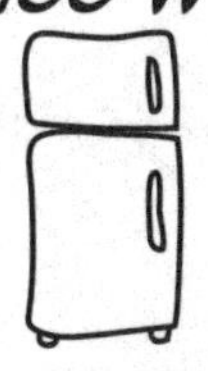

A REFRIGERATOR HAS AN AVERAGE LOAD OF 100 WATTS

250 W

250 WATTS ARE NEEDED TO KEEP YOU MOVING AT ABOUT 25 KM/H ON YOUR E-BIKE

400 W

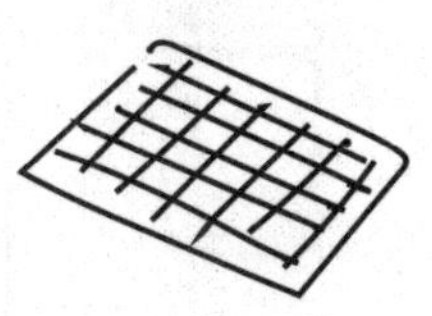

EACH SOLAR MODULE OUTPUTS 400 WATTS AT PEAK

POWER IS MEASURED IN WATTS. 1 HORSEPOWER = 0.75 KW = 750 WATTS = 750 JOULES PER SECOND.

1 KWH OF ENERGY CAN ...

POWER 5–8 KM OF ELECTRIC DRIVING

FUEL 1 DAY OF MANUAL LABOUR

LIGHT A ROOM FOR 60–80 HOURS

COOK A PASTA AND SAUCE MEAL

DO 1 LOAD OF DISHES

BOIL 10 LITRES OF WATER

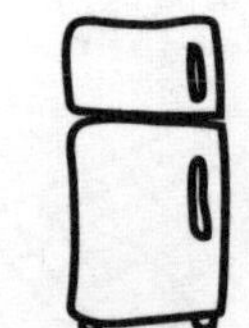

RUN YOUR FRIDGE FOR 6–10 HOURS

TAKE YOU 40–60 KM ON YOUR E-BIKE

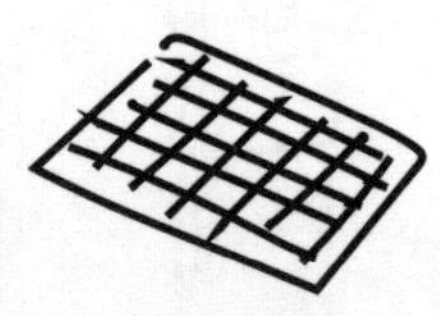

COME FROM 1/2 A DAY'S PRODUCTION OF ONE SOLAR PANEL

ENERGY IS MEASURED IN KILOWATT HOURS.
1 KWH = 3,600,000 JOULES.

Part 1

The Big Picture

1
Rewiring Your Life

At this point, it has become glaringly clear to anyone paying attention that not only is climate change real, it's a crisis. Wildfires are spreading, coral reefs are dying, food systems are stressed and our way of life is threatened. It is long past time to act. The situation is urgent.

That's the bad news. But the good news is that there's something all of us can do about it that will actually make our lives more comfortable and save us a lot of money. The best thing to do to help our planet, our kids' futures, and – it turns out – our wallets, is to make a handful of good decisions over the next ten to twenty years. It's about the next car you own, your next cooktop, what to do when your water heater needs replacing, how to heat your house, and where you buy your electricity from. We just need to plan well and replace any outgoing fossil-fuelled machine with an electric one.

In other words, we need to electrify everything in our lives.

Sometime in the 1970s, the environment was dragged from the realm of scientific evidence to the battleground of partisan politics. Democratic President Jimmy Carter put solar panels on the White House and Republican President Ronald Reagan pulled them down. Environmental action and clean energy have been framed as grim austerity. Self-interested billionaires fund think tanks that continue to attack climate science and spread misinformation. Opportunistic politicians conflate climate solutions with partisan identity politics. I can't change those things.

What I can do is tell you a true story about the near future, because most of it is happening today. We can rewire the whole negative narrative. The data is clear: eliminating polluting fossil fuels will reduce people's energy costs, which will reduce their cost of living. If you electrify the machines in your immediate world, your life will be better. If your whole community does it, tens of millions of dollars a year that would have fled to petrostates, multinationals and offshore investors will stay in the local area. That means more local jobs and local wealth. Your car will accelerate faster. Your pasta water will boil in half the time. Your home will be cosier in winter, and your showers will be just as hot. And even as the temperatures rise, you'll keep cool in the summer, for cheaper.

Electrifying everything is common sense, not a political stance. Both sides of the political spectrum in Australia already know they can win votes by helping consumers buy rooftop solar. They can now add batteries to this agenda, and electric appliances and electric vehicles too.

My motivation is to help us solve the climate crisis as fast as possible, because every fraction of a degree counts. But this Electrify Everything movement could be the start of something bigger – a new social contract, with healthier air, cleaner living and revitalised, local economies. I founded Rewiring Australia in 2020 to accelerate Australia's clean energy transition, focused on policy, regulatory, finance and community solutions. Already we have more than seventy-five groups around Australia affiliated with Rewiring Australia as part of our Electric Community Network. These overlap with the full political spectrum of electorates: 43% Labor, 25% Liberal, 12% National Coalition, 2% Green and 13% independents. In my more hopeful moments, I like to think that as communities electrify, they might get into the habit of working together to solve other problems, using evidence-based solutions and caring about one another.

As an entrepreneur and inventor with a PhD in engineering from MIT, I guarantee that electrification will not stop the economy or make your life uncomfortable. It will mean more jobs, more comfort and more resilience in a world facing climate and trade shocks. It is much more than just an answer to our climate woes; it is the pathway to a more prosperous, and – *if we get it right* – fairer economy.

This transition to clean energy is far simpler than the media would have you think. Basically, it just means not buying any more fossil-fuelled machines and buying electric ones instead. This simple statement reduces to roughly half-a-dozen decisions in your life. These decisions determine about half of the average person's carbon and methane emissions, and are as simple as which vehicle you drive and which appliances you use in your home.

The other half of your emissions come from living in a society that provides schools, healthcare, roads and other shared public infrastructure, including our export industries, which generate income for the national coffers. All these things create carbon emissions, but they don't have to. Eventually we'll electrify just about the whole economy and the country will save money, because machines will be more efficient and energy cheaper. This is what economists like to call "increased productivity". An electrified industry will make clean steel, aluminium, nickel, copper and lithium, which will become electric cars and appliances, for use in electric homes using zero-emissions electricity. It's really that simple.

Why is electricity so much cheaper and efficient? Basically, when you burn fossil fuels to do something, the laws of thermodynamics mean that up to three-quarters of the energy is lost. When you do the very same tasks with electricity, very little energy is lost. Efficiency improvements in the late twentieth century emphasised making fossil fuels slightly less bad – the "reduce, reuse, recycle" slogan – meaning less inefficient. The emphasis now needs to be on electrifying everything and doing away with those inefficiencies altogether. The updated version of this slogan might well be "rewire and recycle". Electrification can give us the good life, more efficiently, at lower cost. Recycling the metals that power electrification will give us a genuine, science-based shot at creating a sustainable or "circular" economy.

So, let's look at what you can do today, right from your kitchen table. The great majority of your emissions come from the infrastructure of your life. These are the big decisions you make from time to time that set the patterns of your energy use, such as how you heat your home, cook your food and get around. These decisions, which you make only occasionally, like when you buy a new car, have a huge impact on how many carbon emissions you contribute to the climate crisis. "Rewiring" is the simple act of electrifying these big energy users and powering them with clean, zero-emissions electricity.

Rewiring your life comes down to five big decisions, representing fewer than ten purchases in your lifetime. Each of these decisions comes up only every ten years or so. These are all the machines and appliances in your life that use energy and generate carbon emissions when you use them.

I want to propose that we start to think of these domestic investment decisions as national "infrastructure", something you more likely associate with roads and sewers and bridges and tunnels. Our team at Rewiring Australia is starting to have some success convincing Treasury officials of this change in definition. Instead of governments investing taxpayer dollars in a few massive energy infrastructure projects such as nuclear power stations, imagine if they invested our tax money in millions of household infrastructure projects? What if every house in the bottom 20% of household income had government-furnished solar and battery systems lowering their cost of living? That would do more to reduce emissions, and faster, than a few nuclear plants or for that matter Snowy 2.0!

There are other emissions associated with things that you buy, such as meat, and infrastructure you rely on, such as roads, but those emissions are not necessarily directly under your control.

If you are a renter or live in strata, you might feel it's impossible to make some of the decisions we are talking about, so we'll also discuss those special circumstances, what you can do, and what policies we should demand of local and federal governments to make electrification work for everyone.

It might offend some people that my answer to decarbonising your life is to buy new things, even if they're electric. You can't wake up tomorrow and create "post-capitalism". But you could wake up tomorrow and make an investment that locks fossil fuels out of your life and locks in lower bills and cleaner air. If enough people follow your lead, we will be almost halfway to solving the climate emergency while creating the demand that solves the other half. We all need things in order to live, and we would all like to live the good life with no apologies. Every home will need a new water heater, stovetop and car at some point in the next two decades. We need to ensure those replacements are not burning more fossil fuels. These everyday decisions, whether they are made tomorrow or in five years' time when your hot water system breaks down, are the impactful ones.

Thinking more broadly, the same principle applies to other material things. I love good tools that work well, appliances that last and cars built to go the distance. We can trade up from fast and unthinking consumerism to more thoughtful materialism. Let's own material things we really value and keep them longer. As Aldo Gucci purportedly said, "Quality is remembered long after price is forgotten" (although I'm recommending long-lasting quality, not Gucci-level luxury).

We all own a lot of stuff, and not necessarily high-quality stuff, as evidenced by the piles left out on council collection days. If you buy good things and take good care of them, you will find your life is less cluttered, produces lower emissions and is cheaper to run. As another famous adage has it: "We aren't rich enough to buy cheap things." But we could all afford to have *good* things if we prioritised creating a future-focused finance system that lets everyone electrify.

For now, let's focus on the emissions outcomes, the things you can directly control – the five big decisions – and on how you can influence the emissions you don't control.

The Five Big Things

You need to electrify the big energy users in your life. These are your transportation, water heating, space heating and cooking. These four things make up about 90% of your daily direct energy use. To make this economical – to save you money – you also need a source, or sources, of clean electricity. For many people, much of this can come from rooftop solar panels and perhaps a battery for storage. Together, the Five Big

Things are:

1. Electrifying your home with solar and batteries
2. Electrifying your driving
3. Electrifying your water heating
4. Electrifying your space heating
5. Electrifying your cooking

The other things in your life that use energy are almost certainly already electric – your coffee machine, fridge, microwave, lightbulbs, computers, tablets and mobiles, televisions, air fryer, stereo, toaster, maybe even your toothbrush.

Some of these changes are easy, and others are a little harder. Personal and cultural factors can play a role. Very few people have an emotional relationship with their water heater, and likely not everyone in your house even knows what it currently runs on. So changing the water heater is for most people an easy change. Some people, however, have a very emotional relationship with their car, whether it be nostalgia, or a preference for their idea of a really cool ride. Similarly, a lot of people have visceral ideas about how food "should" be cooked, and doing things differently can touch on emotional and cultural sore points. We'll take a look at these later in the book – but recall that we've only been driving cars and living in suburbs for around 100 years, and cooking with gas for about the same, yet humans have been living fulfilling lives in vibrant communities, cooking great food, for thousands of years.

My mother grew up in a household that cooked on a coal-fired oven. We embraced gas; now it is time to embrace electrification. Things change. Things get better.

Just to be clear, I am not telling you to go off-grid and become an energy survivalist! We should all stay connected to the grid but we should make as much energy as we can, as locally as we can: on our houses, on our public and commercial buildings, alongside the local railway line and above the car parks where all the EVs will sit. The grid isn't perfectly carbon-free today – far from it – but it gets cleaner every day. This means any electrified asset you buy today is appreciating in a climate sense – it gets cleaner every year, as the grid inevitably moves towards zero emissions.

Tradies will save the world

Every story needs a hero. This book is a love story about saving the world from devastating climate change by decarbonising our lives through electrification. But who are the heroes of this story? You, the reader, are one, but there are others.

The climate story to date has leaned into nerdy heroes. The James Hansens and Al Gores, the Greta Thunbergs and Tim Flannerys. All these people have done an excellent job of highlighting the problem and raising the red flag, but I don't think I'd trust any of them with a screwdriver, and it is now time for screwdrivers. You are a hero, because you are reading this book! And it will be you making the dinner-table decisions that result in the fastest climate actions we can take.

If you follow any of the advice in this book, however, you'll soon be speaking to a tradie to fulfil your electrification dreams. Riding in on an unbridled white ute (soon to be electric), they'll flick their locks in the sunshine as they save you from high energy prices and carbon emissions.

You see, the climate actions that make a big difference in your life do so because you use these things, these machines, every day. You need them to be reliable, installed correctly and maintained properly. All of this turns the traditional climate-hero narrative on its head. The real climate heroes of tomorrow won't be a handful of radicals debating which slogans to use; they will be millions of consumers choosing the right appliances to buy, and then hundreds of thousands of tradies on people's roofs, under people's houses, out behind the garage and under the kitchen bench, making it happen. We won't beat fossil fuels by being angry about the fossil fuel industry, but by putting them out of business. Don't blame the energy system. Change it.

Australian households own around 20 million cars, 12 million stoves, 12 million water heaters, 6 million heating systems and around 4 million solar installations. Replacing these machines, upgrading them, installing them and maintaining them will determine whether we win or lose the climate fight.

Want to be a hero? We need you. Jobs and Skills Australia (JSA), an independent government group that provides advice on current, emerging and future workforce, skills and training needs, estimates that Australia will require 2 million workers in building and engineering trades by 2050 to prepare Australia's energy grid and industrial base for net zero, including some 50,000–75,000 extra electricians. Right now, a lot of tradies "sell against" the electrification program. They show up, you say, "I want a heat pump water heater." They likely say, "Yeah, nah" and pitch you some outdated talkback radio pseudo-facts about "gas is better". That needs to change. We need to invest in our tradies as the critical workforce, *and as the vital point-of-sale people,* of this energy transition. We need to celebrate them as the frontline workers of climate action and have them feel proud that their jobs are the ones that really make a difference.

The not-so-little details

If you don't have solar, you'll need to buy electricity, and we'll look at the cleanest way to do that. To connect that clean electricity to your big loads, you might need three other pieces of kit: a battery, a smart or upgraded electrical panel, and an electric vehicle charger. All these systems will be connected through what the industry calls a "home energy management system" (HEMS), which will make them work efficiently and seamlessly together. At Rewiring Australia we keep it simple and call it a "smart energy device". It's the little computer that tells everything when to turn on and off and manages your electrical "loads".

Importantly, we'll look at how we can finance all these things. So as well as chapters on buying the five pieces of kit, we'll also discuss financing and how we might include everyone in this revolution. How to get all the pieces of kit to work together and to work with the grid. The future of the grid and what we need it to deliver. And we'll answer questions about how we buy electricity today and how we might buy it going forward.

You don't have to read this book straight through; it's a bit of a choose your own adventure. If you want to get straight to the nuts and bolts of a specific thing like solar or induction cooktops, read Part 2 first. To understand the importance of making sure that everyone, including renters, is able to make the big switch to electrification, see Chapters 4 and 5. To be enraged by the rules of electricity regulation and find out how to lobby for change, go to Chapter 12. (Don't skip this one.)

That should pretty much wrap it up, but I'd like to finish with a vision for the future that can be shared. All social change requires an effort in collective storytelling. People – all of us – respond to stories, which is why, along the way, I've shared the stories of people who have already electrified their lives and know how it's done. These are friends and fellow experts who are already a long way down their own electrifying journeys. These stories are about abundance and thriving and succeeding, as we navigate one of the biggest shifts in human history – a shift to a very electric, hopefully quite eclectic, and truly sustainable society.

Everyone will have to take their own electrification journey over the next few decades if we are to cut our carbon emissions all the way to zero. Most of my friends who have done it have enjoyed the ride. Let's start with one of their stories.

2
Electrification Is a Journey

My mate Chris Kerr is happy and optimistic and seems more like a guy you'd meet at the skate ramp than the CEO of a company that is investing heavily in the future of electrification. He's a father of four daughters aged between six and twelve, and his electrification journey started when the eldest (twins) were tiny.

Chris is a busy guy and a self-confessed nerd – or "reincarnated engineer", as he puts it. He likes technology and enjoys data. His initial focus was on what he calls "return on investment", or ROI: he followed the pennies and obsessed about the bills. Quickly he ran into trouble on the home front, however, when his over-optimisation led to a cold shower for his wife. By accommodating his wife's preference for a hot shower, he realised that the electrification journey can also be about ROC: "return on comfort". If we only thought about ROI, we'd all drive smaller cars, shiver in our jumpers in winter and take short, cold showers. But in the real world, we like a little (or a lot) more than just the basics. This is where ROI meets ROC.

Chris's daughters had different motivations again for electrification.

Like most kids today, they were very aware of climate change and the need to change our ways – for their future. They drove the ROE agenda in the household: "return on environment".

I love this ROI, ROC and ROE framing. It very simply expresses the value of electrification and the opportunities it brings. Being good for the planet no longer needs to conflict with your finances or your health and well-being. Wherever you start on your electrification journey, it can improve your life, the lives of your loved ones, and even the lives of the magnificent array of creatures we share the planet with.

For early adopters such as Chris, this household electrification project was not just a journey but an odyssey, filled with mistakes, missteps and misadventures. As with any adventure, the hardships resulted in valuable lessons, good stories and a sense of contentment once the mountain had been climbed. Now that people such as Chris, and the other early adopters in this book, have spent the time and made the mistakes, the journey will be far easier (and cheaper) for the rest of us.

In Chris's case, his journey began with an electric bike. He had a fancy Audi, but the first time he tried an electric bicycle, he found that the commute across northern Sydney was so much fun and so fast – and free! – that he left his expensive car sitting in his driveway for six months, largely unused. This has been my experience of electric bikes too, and others have reported the same. It is for this reason that I'll dedicate a little space in the book to the humble bicycle. Or, rather, the not-so-humble, car-replacing, life-affirming, community-enhancing electric bicycle.

Chris's electrification journey has been a bit more complicated than the simple "five things" framework I gave you earlier. Yet it isn't in conflict with that simple advice. The five things are the most impactful things you can do in terms of the measured emissions of most Australian households. But plenty of other little things can also be part of the journey. These are the cherries and whipped cream after the meat and three veg of electrifying your home.

Let's see how Chris's electrification journey played out.

Chris's electric bike was the gateway drug – it was just a better way to get around. He got to work faster, with the same amount of effort, less sweating, less traffic and more time in the sunshine, and it was a far cheaper commute.

Then he got his first of the Five Big Things – solar – spurred by an electricity bill shock. He saw the $1500 quarterly bill and realised he had to do something about it, so in 2019 he installed 8 kW of solar. An 8 kW system makes about 25 kWh of electricity per day – nearly double the average Australian household's current electricity use (you can find a rundown of kW, kWh and other electricity terminology in the FAQs section of this book). Chris was now producing more electricity than he needed, which led to the next step.

Once he had cheap solar electricity coming in, Chris got stung by the energy companies when they changed their feed-in tariffs – they were paying him less for the excess electricity he was feeding back into the grid. This made him realise he'd be better off if he increased his load – that is, if he used more of his own electricity by adding more electric things. This was the second of his big five. He bought an electric water heater with a 3.7 kW element and a Solar Edge load controller.

Even with the electric water heater, he still wasn't using all his solar production. In a classic case of what might be called "dad maths", Chris invested in an electric hot tub. Any time he was producing excess solar, he could dump the heat into his spa and get a high ROC along with his ROI. I have used this exact same logic and installed my own heat pump hot tub.

A well-insulated hot tub will use about 4–7 kWh per day, or roughly the output of three to five solar panels. Some well-meaning climate activists might scoff at this luxurious idea, but I believe with some conviction that we aren't going to solve climate change if it doesn't improve our lives. Water is a very effective battery, in that it stores energy, particularly if well insulated, so you can dump your excess load into the hot-tub battery and use it later when you want to have a spa with the family.

Chris's next investment was an induction cooktop, another of the big five. This quickly led to an electric barbecue as the family decided that electric cooking tasted better – no fuel on their food. I've also found that to be true; that faint taste of unburnt hydrocarbons goes away, and the air in the kitchen smells and feels cleaner. Chris's family now runs an outdoor kitchen most nights of the year, with electric air fryers and an electric barbecue on the deck.

The next step was electrifying the lawn. I spent years learning to speak the language of small carburettors, and like many Australians I found that language was mostly expletives. Not only is cutting lawns bloody miserable, but it is made more miserable by the noise and fumes of petrol mowers and obnoxiously loud weed whackers. Chris found it easy, and then pleasurable, to switch to an electric mower. Like most electrification, the electric alternative was simply a far superior product. Someone out there will probably complain that this isn't yet true for the family with two acres of grass and a ride-on lawnmower, but trust me, I've seen the future, and the ride-on electrics will also be quieter, cheaper to operate and better than the current machines.

All of this load-growth (the technical term for adding electricity demand by adding electric machines) still wasn't quite consuming all of Chris's solar, and it wasn't consuming it at the right times of day. He could see he was still paying high electricity rates on the shoulders of the day (in the early morning and late evening, when demand is highest and the sun is lowest). It was during Covid, and the family couldn't take a planned vacation, so, using some more of his dad maths, he pushed the holiday money into purchasing a 20 kWh battery – plenty to power his house and his electric toys.

The next investment was the electric vehicle. Chris had been waiting for an electric car that could also occasionally provide power to his house, since an electric car has a battery, and a battery is energy storage. This emerging technology is sometimes called vehicle-to-grid, or vehicle-to-load. The car that satisfied his personal needs was the Hyundai IONIQ 5. It turns out that per kilowatt hour, the battery in the car cost the same as he'd paid for the battery on the side of the house – around $1000 per kWh – but in the car's case, it came with four wheels and could accelerate like a Melbourne Cup winner. In just a couple of years, Chris has already put 70,000 kilometres on the car, and the only maintenance has been one change of tyres. He estimates that he has spent considerably less than $1000 in total on charging the vehicle, as he mostly "fills up" with electricity from his house.

At this point, the only thing left to electrify was the pool heater. Not to heat the pool year-round, but to act as an additional soak for his very comfortable, very low-cost rooftop solar. Why not extend the swimming season for that little bit extra ROC and ROI?

Chris still has a diesel 4WD in the driveway. Eventually he will replace it with something electric or hybrid, but for now it barely gets used. I have a diesel HiAce fitted out for camping in my driveway – who am I to criticise? My van, like Chris's 4WD, gets used once every month or three for fun family trips. Depending on who is driving furthest on a given day – Chris or his wife – that person uses the electric IONIQ. Most other trips are done by electric bike. Chris has just used some more dad maths to buy a new electric mountain bike, something he justified because the household was able to survive a bunch of grid outages thanks to the batteries and the car. Why not invest in more of this high-quality life?

Chris talks passionately about the new smell of his life, or rather the lack of smell. No diesel fumes, no gas smell in the kitchen. He is never going back. His initially resistant wife now has a never-go-back policy, too. She anticipated it would be a pain in the arse to plug the car in. Now she can't imagine driving to the servo to buy petrol. His girls don't have a never-go-back policy, because they don't need one: they don't really remember what life was like before electrification, the same way my mother barely remembers riding in her grandfather's horse and sulky. In the same way, my children will never remember cooking with gas.

Chris is now calculating when it will make sense to upgrade and increase the size of his solar system, probably a good thing to think about given his four girls will be driving age soon.

And Chris's family aren't the only ones enjoying the ROC. There's a family of possums who love the warm spot under the solar panels in the winter.

Every family needs an electrification plan

Chris's journey, like my own, and like many of the people I talked to in creating this book, began with a bit of a plan. For many it starts with solar panels, for some with the car, for others the water heater, or occasionally the whole project is precipitated by a move to a new house.

I already had an electric car when I recently moved house. The house has an old, tiny and insufficient solar system and came with an in-line gas heater and a gas stove. I quickly retrofitted my own electric cooktop, so now I need better solar, an electric water heater and a better car charger. I'm going to do those three things at once, as there will be a bit of a discount doing the work in bulk and financing it all together.

I'm waiting for the VW electric van before I replace my old Toyota HiAce. I will electrify a vintage car with my kids so that they understand the components of this new electric world and have something with personality to drive. For fun, I'm electrifying my two-stroke 1966 Vespa with a sidecar, and then I'm electrifying my Zodiac inflatable dinghy. The battery in the boat will be part of my energy storage strategy. As my wife knows only too well, the boat just sits on a trailer next to the garage most of the time. With a 60 kWh battery, it will be a fabulous backup for our solar energy, giving me an excuse to keep it parked there.

Your plan will be different. And like me, you'll probably change your plan along the way, or discover that you like how it is going and decide to do it faster. I'm moving pretty fast – I'm lucky I can afford to do so. No doubt we need more solutions for renters, solutions for strata, and solutions for people on low and fixed incomes who would benefit from the savings involved in going electric but whose access to credit and finance are severely limited. Suffice to say, Australia does not yet make it easy for everyone to electrify and decarbonise. This is why I spend a lot of my time agitating for federal and state policies that will help every household, regardless of income, succeed on this journey. We need to agitate for change and support politicians in making all these things easier for everyone.

How do you create your own plan? Each situation is unique. If you're a renter, there are only so many things you can do, but you can get an induction cooktop and an EV to start. If you're a homeowner, it makes sense to start with the appliances you use most frequently and can afford to replace, such as an induction cooktop or split system air conditioner. Ultimately, you'll save the most money and have the most impact if you install solar, because our rooftop solar in Australia is the cheapest energy in the world, and it will power all your other appliances and EVs. A home energy management system (HEMS) will tie it all together and help you to power your home most efficiently.

Whatever you do, when one of your fossil-fuelled appliances dies, replace it with the electric equivalent. You'll save money in the long run, your air will be cleaner, and you'll have a better machine.

Take a piece of paper and write down these sentences, and you'll have the start of your plan.

1. I think I will be buying a new car in ________. I'd like it to be an electric ________.
2. I believe my water heater is ________ years old. That means in ________ years' time I will need to replace it with an electric resistance or, even better, electric heat pump water heater.
3. I can improve the respiratory health of my family and make cooking cleaner and easier with an induction cooktop. We'll be doing a renovation in ________ years; I can do it then, or maybe my partner and I can gift it to ourselves this Christmas.
4. My space heaters are ________ years old. I have used an online calculator and found that I could save $________ per year on heat. I will replace my heating system in the year ________.
5. I know I could be saving immediately on my electricity bills. I will buy rooftop solar and perhaps a battery as well in the year ________.

6. I'll start saving now and investigate finance options and government incentives to enable these purchases.

And there it is: you have a rough electrification plan. Being cheeky, you might add this one:

7. I'm going to borrow a friend's electric bike, and if I like it, I will ask for one for my next birthday.

3
Energy, Energy Bills, Electrification and You

The principal advice of this book is to electrify everything. The promise of the book is that it will save you money (and save the climate). Does that mean it will lower your electricity bill? Probably not, because you will use more electricity. Will it lower the total cost of all your energy bills? Almost certainly, because electricity is much cheaper to use than petrol and diesel and very often cheaper than gas. We need to think about our energy purchases in a new way.

You almost certainly don't think much about all the ways you use energy in your life. You probably don't know exactly how much money you spend on energy, or all the different ways you spend money on energy. Or maybe you know exactly how much you spend on some of your bills, and it horrifies you. Unfortunately, it isn't in any corporation's interest to make you energy-literate, as the companies that sell you fossil fuels – whether gas, LPG, petrol, diesel or coal- or gas-fired electricity – profit from people's ignorance.

Many of us have had the experience of subscribing to something like

Netflix or Stan, only to forget we have the subscription and then find out, months or years later, that we have been paying for something we haven't been using. By default, we've all had a subscription to fossil fuels all our lives, and the subscription companies aren't going to send you a reminder that you could be doing things differently. You are going to have to cancel those subscriptions yourself. My friend Leah Stokes, the wonderful climate policy nerd, calls it D-day. The D is for disconnection – the day you have finally disconnected from gas networks and petrol networks and are a zero-emissions household.

Later in the book, we'll look in more detail at how the current energy market works and how we can make it fairer. In the meantime, this quick primer is to give you the minimum knowledge you need to understand how you pay for energy today, the costs of different energy sources to your household, and how each one contributes to carbon emissions. Armed with that information, the rest of the book will help you to pay less for energy tomorrow.

How energy comes to you

Energy services are so ubiquitous in the modern world that we have forgotten where energy comes from. When I first moved to Wollongong with my (American) wife, she observed a train loaded with black stuff going under a rail bridge and asked, "What is that?" I told her it was coal. Later that week she saw a nice black rock on the ground during a hike, and I had to inform her that Australia is literally made of the stuff. Even though coal creates a lot of our electricity, many of us don't know what it looks like – or perhaps didn't until Scott Morrison took a chunk of it into parliament.

You can't see natural gas until you light it on your stove, and the infrastructure to deliver it has become invisible to all but those few who bury the pipelines. You swipe your credit card or tap your phone, stick a nozzle in the side of your car, then drive away a minute later. You never see the 40–60 litres of fuel that fills your car. Your car works reliably and powerfully, as if by magic, but it is actually burning the equivalent of 150 Coke cans of petrol or diesel with every tank. If you expressed fuel economy in kilometres per Coke can, you would only get about 4 kilometres per can.

You catch a whiff of them, but no one ever really sees our energy sources. Because we have made them invisible, and because the greenhouse gases they cause are invisible, we ignore them all. Let's change that, by taking a look at what is really going on behind your energy bills.

Your energy bills today

From statistics and surveys collected by the Australian government, we can draw a picture of energy use by a typical fossil-fuelled household, as highlighted on the left bar of Figure 3.1.

You buy energy directly, most of it fossil fuels, in four or five ways: when you buy petrol or diesel for your car; when you pay for electricity to be delivered to your home; when you pay for gas to be piped in; and when you buy propane or bottled gas for a holiday house or a barbecue. You might also have bought solar cells.

Obviously (or perhaps not so obviously), you are also buying energy when you buy furniture or food; it is embodied in the product. But here we are only considering the energy purchases that a household controls directly, not the indirect or "embodied" energy.

Figure 3.1 summarises what these direct uses of energy look like for the "typical" Australian household. The average Australian household has 2.6 people in it and 1.8 cars, as well as electricity and a gas connection.* No one has 60% of a kid or drives 80% of a car, so it is obvious that the "average" does not describe a real household, but it is useful for thinking about the problem. When we talk about a "typical" household, you can imagine a household of three people, with two petrol cars, gas heating and gas cooking.

Of course, there is a lot of individual variation. If you have a big or young family, you probably use much more energy. If you are a retired couple whose children have left home, likely much less. Nevertheless, it is useful to think about this average, because when it comes to energy-use behaviour, we are more alike than we are different.

Today, our typical Australian fossil-fuelled household uses around 120 kWh of energy per day. Of that, around 45%, or nearly half, is for vehicle fuels. Around 10% is for space heating, around 8% is water heating, and a couple of percentage points are cooking. All told, about 10% of the energy used by a household is gas or propane, used for heating and cooking.

Fourteen percent of the household's energy is used for electricity (some of which heats the water and space). Not typically shown in statistics, but something I want to emphasise here, is the 17% of the household's energy that is wasted every day by using coal or gas to generate the electricity. This means that around 30% of today's "typical" household load is "electric", but it is inefficiently done. When we move to solar and wind, the great majority of this waste will disappear.

We can use this overview of energy use and combine it with typical prices of the fuels you buy to translate that energy use into energy costs.

---

* Australian Bureau of Statistics, Household and Family Projections, 28 June 2024, https://www.abs.gov.au/statistics/people/population/household-and-family-projections-australia/latest-release#households

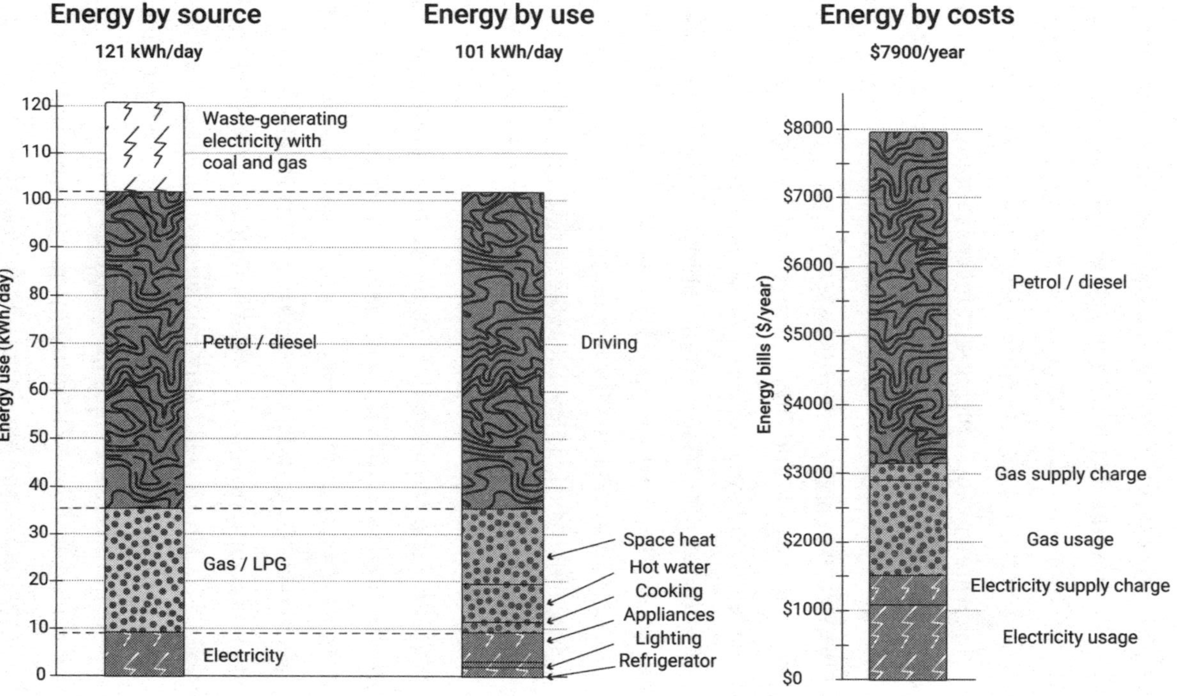

Figure 3.1

Energy use for all activities of a typical fossil-fuelled Australian household. The first bar shows usage by energy source, including the wasted energy generating electricity with fossil fuels. The second bar breaks down the use of energy by activity – driving being the most energy intensive thing we do. The third bar breaks down the cost of energy for the typical household.

Again, you can see that the dominant cost for the Australian household is petrol and diesel, which is why electric vehicles can be such an economic boon for Australia. The cost of electricity and gas has two components: the use charges, for how much energy you use, and the service charge, which is the daily fee just for being connected. For the typical household, the daily service charge is around a dollar for electricity and an additional dollar for gas. If you aren't using much gas, it makes sense to cancel that subscription by electrifying everything, which will save you around $300 per year on the service charge alone, in addition to the savings you'll make by running your new appliances on electricity.

To compare your current and future household energy bills, we'll need to know the relative prices of energy for doing different things with different energy sources. For this we can look at Figure 3.2. In Australia, most electricity plans let you buy electricity from the grid at 25–40 cents per kWh. If you are lucky enough to have rooftop solar, depending on where you live, how much you paid and the interest rate on your loan, you are probably getting some of your electricity for between 4 and 8 cents per kWh.

Petroleum-based fuels are wickedly expensive by comparison. An equivalent kilowatt hour's worth of driving a petrol or diesel car will cost you $1–1.20. Gas is typically used for heat in the home, whether to heat water, air or food. It typically costs 25–50 cents per kWh of heat equivalent. Grid electricity powering a resistance heater – such as an old-fashioned electric cooktop or radiator – will likely cost about the same or a little more. If, however, you can use a heat pump for your water, or a heat pump for your space heat, the incredible efficiency of that heat pump means that effectively you will be paying only 10–20 cents per kWh equivalent. (I'll explain what a heat pump is, and how to get one, later in the book.)

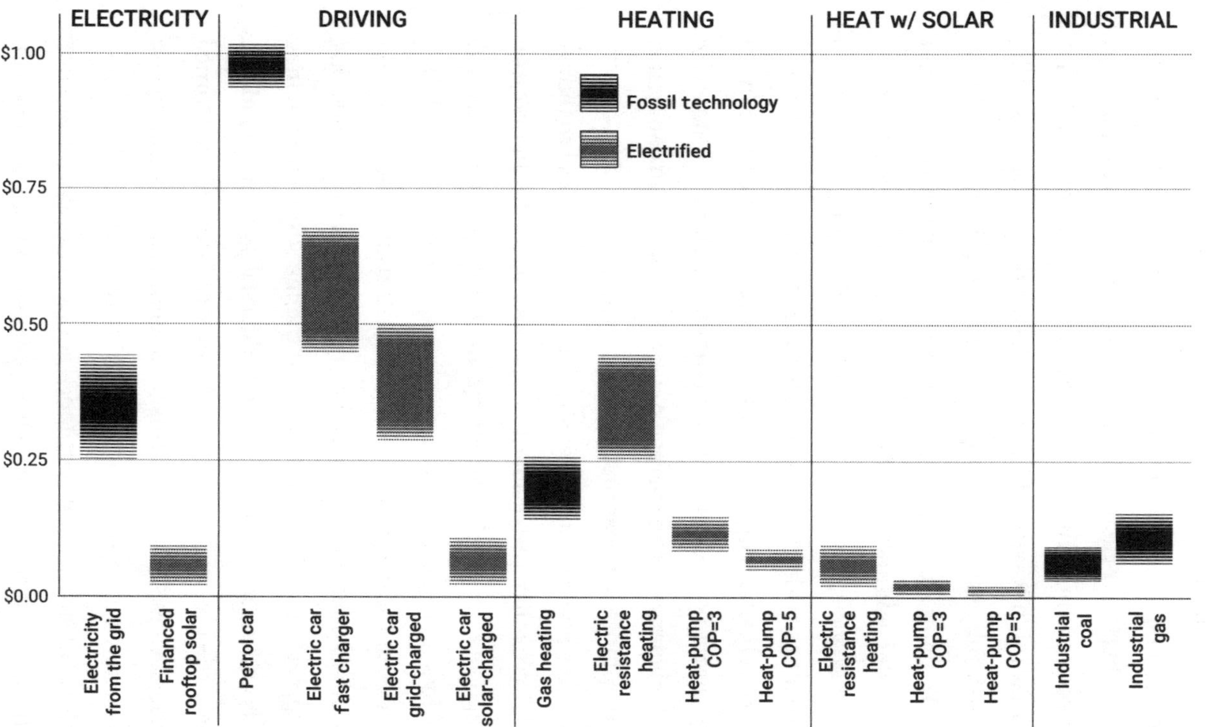

Figure 3.2

Comparative costs of 1 kWh of useful energy for different activities. Electricity: costs per unit of grid electricity versus financed rooftop solar. Driving: approximate costs of driving with petrol versus electric vehicles charged in various ways. Heating: costs of 1 kWh of heat from burning gas, from electric resistance and from heat pumps. Heating with solar: same heating systems, but powered with rooftop solar. Industry: shows how cheap industrial energy is, and why it is challenging to decarbonise industry today.

Figure 3.2 is extraordinary when you think about it. For the things everyone does every day – using electricity, driving cars, heating homes, heating water – this chart clearly shows that Australia has a simple and easy pathway to zero emissions and to a far cheaper energy system that doesn't rely on foreign fossil fuels or multinational corporations. Why aren't we doing it full speed, then? Well, I actually think that this is partially consumers' fault. We aren't broadly aware of the opportunities, and we're not clearly demanding them from the government, from the energy sector, from our service providers or even from ourselves.

One extraordinary thing to note in Figure 3.2 is that if we use renewables to completely electrify this "typical" household, it can run on about 40% of the energy it used to, without doing anything traditionally thought of as "efficiency". The average Australian household could save $4000, $5000 or more per year just by electrifying. In the fifth column, I have shown the wholesale energy cost of gas, coal and diesel to industry. It is much lower than the residential cost because it doesn't require the same complicated network for delivery, and the price is negotiated in bulk. This is why some newspapers – I'm thinking specifically of *The Australian* and the *Australian Financial Review* – still argue that the energy transition is too expensive. But that's because they aren't comparing apples with apples. Electrification with renewables will save us money on households and vehicles, and lots of it. Industrial solar and wind will eventually beat the low wholesale cost of industrial fuels as well, probably within the decade. This is why this book emphasises the domestic economy opportunity of electrifying our castles and cars.

There is a lot of misinformation around, trying to convince us that cooking with electricity is not as good, that electric cars aren't "manly" enough, or that climate change simply doesn't exist. Meanwhile, the environmental left is drowning us in a different kind of misinformation – guilt about which type of t-shirt we wear or what type of coffee cup we use. In this humble little graph, however, is the underlying energy revolution that, whether gas companies like it or not, is inevitable – because it is economically irresistible, and because it is the answer to our climate conundrum.

Many gas companies will direct you to websites suggesting that electrification will cost you more – but they cynically use traditional resistive electric water heaters for their comparisons, not the far more efficient modern heat pumps. They also don't factor in the cost savings of using rooftop solar. And, because they really don't want you to know, they rarely if ever publish their methodology, so it is hard to assess or refute their claims. These kinds of comparison websites are designed very badly or very cynically, depending on what you believe their motivation to be.

Many people associate me, and Rewiring Australia, with home electrification, but I'm actually more of an industrial guy; my career started in steel mills. I emphasise household electrification not because it is the only important thing, but because it will save us money *now*. Green steel is coming, and green industry too, but we are a decade away from them working cost-effectively at scale. In the meantime, let's get on with climate action now and electrify our homes!

Electrification is the biggest efficiency

Figure 3.3 lets us look in detail at the energy savings involved in electrification, and at the money savings, both of which should tantalise every household and the nation at large.

We've already seen that of the 120 kWh per day that powers the average Australian fossil-fuelled household, about 60% of the energy is wasted. The biggest slice by far of our energy use is vehicle fuel – even more of which is wasted! Vehicle fuels make up just over half of total household energy use, which explains why one of the chief strategies for decarbonising Australia is switching to electric vehicles. A car engine wastes 80% of its fuel as heat, which is why you can burn your fingers on the engine block. That means only about 20% of the energy content of petrol or diesel turns your car's wheels and moves you along. An electric motor, by comparison, loses only 5–10% (and you're much less likely to burn your fingers). That is why a similarly sized electric car powered by renewable energy uses just one-quarter of the energy of a petrol vehicle to go the same distance. Another reasonable chunk of an average household's daily energy use, around 6–7 kWh, is spent heating water for showers and laundry. Space heating similarly uses quite a lot of energy, averaging about 9–10 kWh per day. If this number looks low to you, you probably live in Victoria, the ACT or Tasmania, where heating costs are higher. If it looks high, you likely live north of Sydney or on the west coast.

Given all these factors, you can see why the all-electric home is the true efficiency we have been looking for.

The only pieces of this pie that can be made substantially smaller through traditional efficiency measures are space heating and cooling, but on retrofitted houses this can be quite expensive for limited gains. You can insulate under the floor and in the ceiling, double glaze the windows and seal gaps around doorways and windows, but those efficiency measures will only apply to 15–20 kWh of your daily 120 kWh of energy use. Every bit makes a difference, and it can be critical in some homes, but it isn't always a good economic choice. Sometimes, adding more solar and a bigger heat pump will be cheaper than hunting down every air leak and trying to create a perfect "thermal envelope".

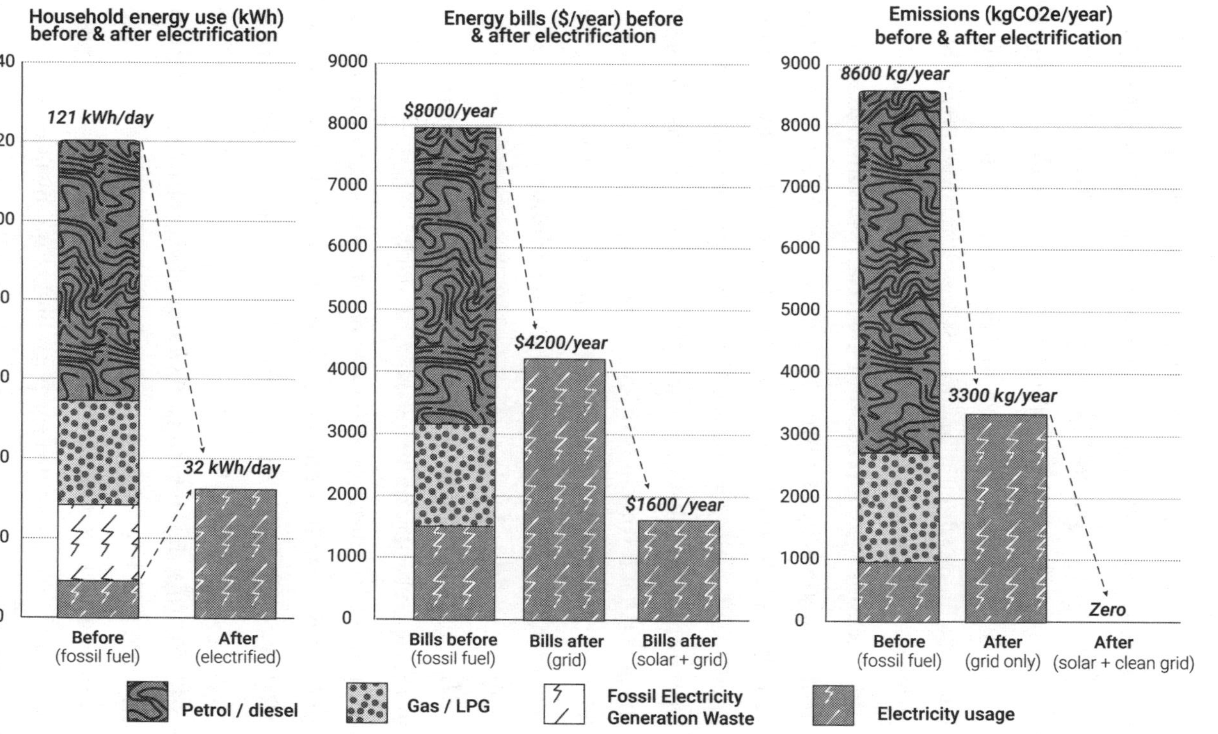

Figure 3.3

Before and after electrification. (a) Energy. (b) Bills. (c) Emissions. (a) Shows usage by energy source and the extraordinary efficiency win of going all-electric. (b) Shows electrification lowering annual bills. (c) Shows emissions going to zero through electrification.

Incredibly, if you just follow the advice in this book, you will eliminate about 60% of your energy use, and your household will go from 120 kWh per day to around 40 kWh per day. This is the seemingly magical efficiency of "electrifying everything" using renewable electricity.

As we saw earlier, your electricity bill will not necessarily be lower after you electrify. Your electricity use will go up, by as much as 300%, so your bill may be higher depending on how much solar you have, how much you drive and where you charge your car. But you will be spending much less on other energy sources. Your electricity bill may go up, but your total cost of energy will go down.

Figure 3.4 shows this dramatic change in dollars. The first column shows a typical fossil-fuelled household today, spending around $5000 per year on petrol and roughly $1500 each on electricity and gas. The second column shows that if you electrify everything and run your home off the electricity grid, you'll spend about half as much on your total energy use, and now you'll have to deal with just one subscription – the electricity. That amount can be halved again if you get about half your energy from your own rooftop solar. It can be further halved if you add a battery so that you can keep using your own solar after dark.

Note that these are "running costs" or operating costs, and don't include the upfront costs of installing solar power or any other items. But as Figure 3.5 teases, the economics still work with those upfront costs included.

Figure 3.4, of course, doesn't tell the full story, because to benefit from these savings, you first have to buy the goods – the electric vehicles, the solar, the batteries. In Figure 3.5, we calculate the total costs of ownership for the lifetime of these subscriptions, including the financing cost of the electric machinery.

Figure 3.4

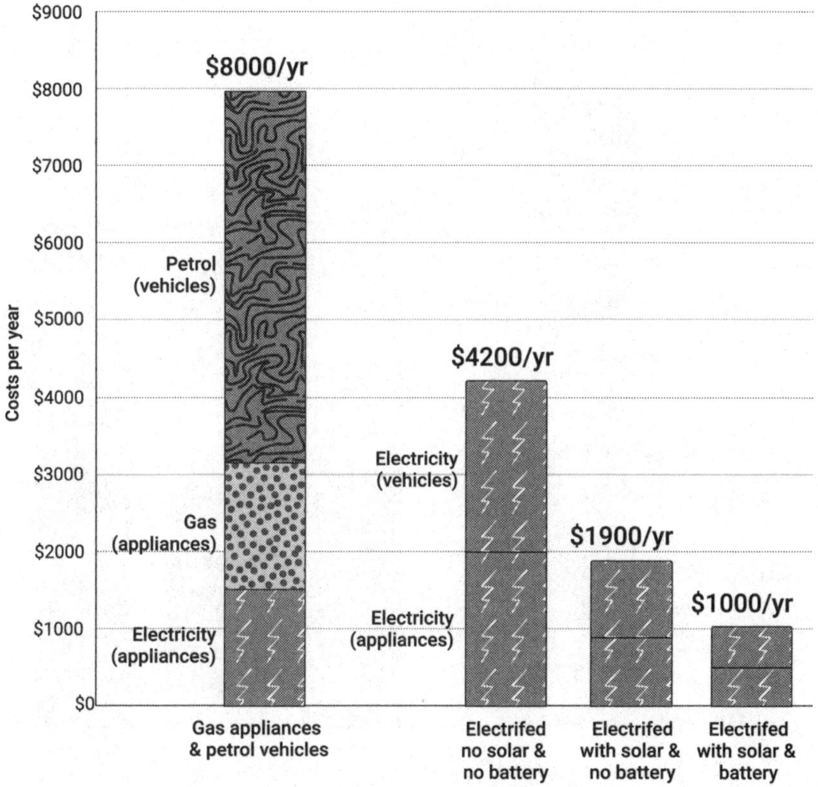

Energy bills per year. The first bar shows a fossil-fuelled home, the second an electrified home using only grid electricity, the third electrified using solar and the grid, and the fourth using solar, a battery and the grid.

This model includes the upfront costs of owning electric cars, water heaters, cooktops, solar and batteries. Yes, electric cars cost more than petrol ones (for now). Yes, a heat pump water heater costs more than a gas one. But as this graph makes clear, the lower cost of operating an electrified system every year adds up, and the investment pays off, as we'll see in detail in subsequent chapters.

Figure 3.5

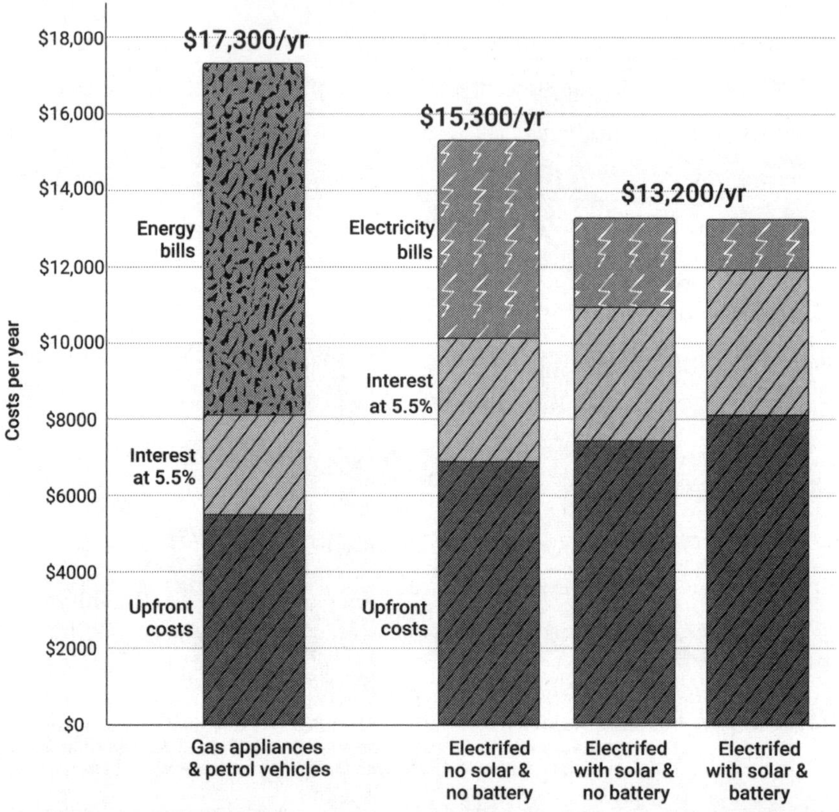

Energy and upfront costs per year. The upfront costs and interest are the total price split over fifteen years. The first bar shows a fossil-fuelled home. The second bar is an electrified home using only grid electricity. The third bar is electrified using solar and the grid. The fourth bar shows a home using solar, a battery and the grid.

All of this is good news. Succeeding at climate action does not need to cost you money anymore. Quite the opposite: it will likely save you money. The bigger threat now is that we don't make it possible for all Australians to afford the transition, whether because they don't have cash, don't have credit, or their cost of finance is too high. This inequity will truly supercharge the climate culture wars. But if we get the incentives right and make these choices easy to access for as many people as possible, then we can all afford this better life. And to be clear, when I say incentives, I don't mean subsidies, a word that riles many Aussies. I mean incentives in the structure of our tax code, our regulatory environment, and even how we apply the rules of banking to financing everyone. Later in the book, we'll look in more detail at how we can achieve this.

When should I electrify?

When is the right moment to do all these electrical upgrades of the machines in your life? You don't have to run out and replace everything right away, unless you're flush with cash and can afford to get to carbon-zero by next month. For the rest of us, we should be replacing things when they die, as we would anyway, and just replace them with zero-emissions electric things. The average car lasts around twenty years. The average water heater, fifteen years. The average cooktop, maybe twelve years. If we replace them with electric machines when they fail, we'll be at zero emissions in twenty years. Simple.

It's therefore important to think about every purchasing opportunity: will the purchase move you closer to zero emissions, or will it lock in another ten or twenty years of carbon emissions? Whether we're buying cars, appliances, power plants or steel mills, we need every purchasing decision from now on to be zero-emissions and all-electric if we are to have any hope of avoiding 2 degrees Celsius of global warming.

For many people, the obvious opportunity to upgrade a dirty fossil-fulled machine to an electric one is when it fails. You don't have spare cash, you don't want to go into debt, the kids are hungry, the dog is sick, and frankly you don't have time to worry about anything beyond this week. But in one of those busy weeks – and we all have them – your hot water heater will fail, or the old gas stove will finally stop working, or the shitbox in the driveway will refuse to budge. When one of these moments hits you on a random Thursday, and you need to replace a fossil-fuelled machine quickly, you'll be more likely to electrify if you've thought about it in advance. Do a tiny bit of research – reading this book counts! – know your options and be ready to electrify that thing when it goes arse-up.

Another good time to electrify is when you renovate or move house. When you buy or rent a new home, or when you're renovating, consider what you can electrify. Similarly, if you get a promotion or have some other big change in your financial situation, it can be a great time to electrify at least something. Refinancing? Another great time to do some electric renovations.

4
Financing Your Electrification Journey

I hear you say, "This guy Saul, he lost me at needing to spend $25,000 to replace all the things in my life that run on gas." Yeah, that would be a lot of money to spend upfront and all at once. But I want people to think about it differently. In the next fifteen to twenty years, there is about a 99% chance you will buy a car, a water heater, a stovetop and a new heating system. You'll probably buy solar, too, whether it's to replace an ageing system or to install a brand-new one. All I'm really saying is that when you do need to buy new machines in the future, buy electric ones, and be ready for it financially.

In the long run, replacing your machines with electric ones will save you money. In the past, when you bought a car, you had to fill it at the petrol station on the way home from the dealership. That was your first instalment in an expensive subscription to fossil fuels that followed that car everywhere. When you buy an electric car, it is more expensive upfront, but it saves you money every time you "fill the tank".

Electric machines can save you money, but there is a problem. Not everyone has access to finance, and not everyone wants to go into debt. If we're going to electrify everything, as I've said, we need to make the money we're investing work for every home in Australia – not just the wealthy, but everyone. Most homes will need financing that enables them to make the upfront purchasing decisions that will permanently lower their cost of living through savings on their energy bills. This could be public or private finance, but the facilitation of that finance is crucial so that all homes have fair access.

Swapping fuels for finance

Electrification can be thought of as swapping fuels for finance. Fossil-fuelled machines often have a lower upfront capital cost, but a much higher operational cost. For example, an electric car might cost $5000 to $20,000 more than a similar petrol car (although prices are reaching parity). If that extra upfront cost can be financed over the lifetime of the vehicle, the cost of ownership will often be lower than the refuelling costs of a petrol vehicle. Your loan repayments each month, in other words, will probably be less than you would have spent on fuel each month if you'd bought a petrol car – you have swapped fuels for finance.

Crucial to this picture is making sure that the finance terms have similar timeframes as the lifetime of the asset or its warranties. As an example, solar panels often come with a warranty of thirty years. If they are financed over the same period – that is, if you have thirty years to repay the loan you took out to install them – the electricity they produce comes in at a financed cost less than a quarter of the cost of grid electricity. However, if the repayment term is only three years, you will need to pay the loan off much more quickly. The household may need to take a cost-of-living hit for three years, in order to unlock savings later on – and taking a cost-of-living hit is not possible for many people.

The basics: the cost of money

It might feel basic to many readers, but I think it is useful to go over the basics of the cost of finance. If you already know everything you need to know about money, skip to the next chapter. If you want to read a couple of thought-provoking books on finance and interest, I recommend *The Price of Time* by Edward Chancellor and *Debt: The First 5000 Years* by David Graeber.

Figure 4.1 shows the multiplying cost of interest as the term of a loan increases. A simple home loan often has a term of twenty years. If you have an interest rate of 8% and repay the loan over twenty years, you will end up paying back double what you borrowed. God forbid you pay 12% over twenty-five years; in that case, you would pay triple.

A typical car loan has a term of five to seven years. If you take out a six-year car loan at an interest rate of 12%, you'll end up paying about 1.5 times what you borrowed. A typical loan on rooftop solar is ten to twenty-five years. A typical battery might be financed over five to ten years. By repaying a loan more quickly, you can minimise the cost of this borrowed money, but your weekly or monthly repayments will be higher. Alternatively, you could avoid taking out a loan altogether if you can afford to buy these assets in cash, but then you have to factor in the opportunity cost of that cash not earning you money through some other investment.

My friend Fred Hopley, who you'll hear more about in Chapter 11, funded his electrification journey with savings, and considers it an excellent investment. Whenever Fred is talking to homeowners considering solar, he lays down a challenge: if they have a better place to put their money, he'd love to hear about it! Investing in electrification delivers a 20–25% return with almost no risk, and there's growing evidence that it increases the value of your house as well. Fred's return on the combined solar/battery system is 10–15% a year and climbing. If your situation is

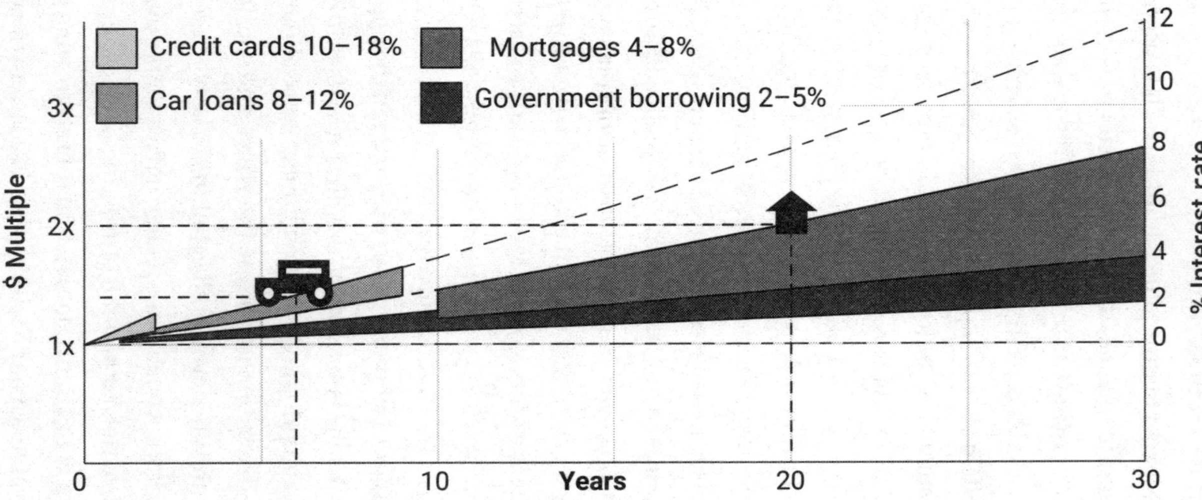

Figure 4.1

The cost of money. On a six-year car loan at 12%, a car costs nearly 1.5 times as much. On a twenty-year home loan at 8%, a home costs nearly twice as much. The cheapest money in the country is the government's rate of borrowing, followed by mortgage rates.

similar to Fred's, you will likely be cashflow positive immediately.

It is worth looking in a bit more detail at examples of the costs of electrification, and the logical first example is solar. Most solar now comes with a warranty of twenty to thirty years; some companies even warranty solar for forty years. Most solar installs on households or businesses cost around $1 per watt of "nameplate" capacity. Nameplate capacity is the system's capacity at noon on a sunny day. To translate this into real-world performance, we use a number called the capacity factor. Typical capacity factors for rooftop solar in Australia are 0.14–0.19. This means that if you have a 1 kW module, it will generate 1 kW at noon, but over the course of the day it will average 0.14–0.19 of that amount, or 140–190 watts, which is about 3.4–4.6 kWh per day per kW of installed solar system.

The further north and the sunnier your climate, the higher your capacity factor will be. If trees shade your roof, or your roof is at an inconvenient angle, the number will be lower. In Figure 4.2, I have assumed a capacity factor of 0.16 and a product lifetime of twenty years. For additional accuracy, I have assumed that about 0.5% of the performance is lost every year as the modules get old, dusty and scratched.

The left axis of Figure 4.2 is the cost in cents per kilowatt hour. The bottom axis is the price you pay in dollars per watt for the installation. If you pay $10,000 for 10,000 watts (a 10 kW system) then you are paying the average price in Australia of around $1 per watt. If you are lucky enough to pay upfront – which means you are paying zero interest – and the system lasts you twenty years, you can see the effective cost per kilowatt hour is around 3 or 4 cents. Even if you are paying 12% interest, the cost is going to average out at less than 10 cents per kilowatt hour.

If you don't use all of your solar each day, you will be paying more for each kilowatt hour you use. For example, if you have no battery and can

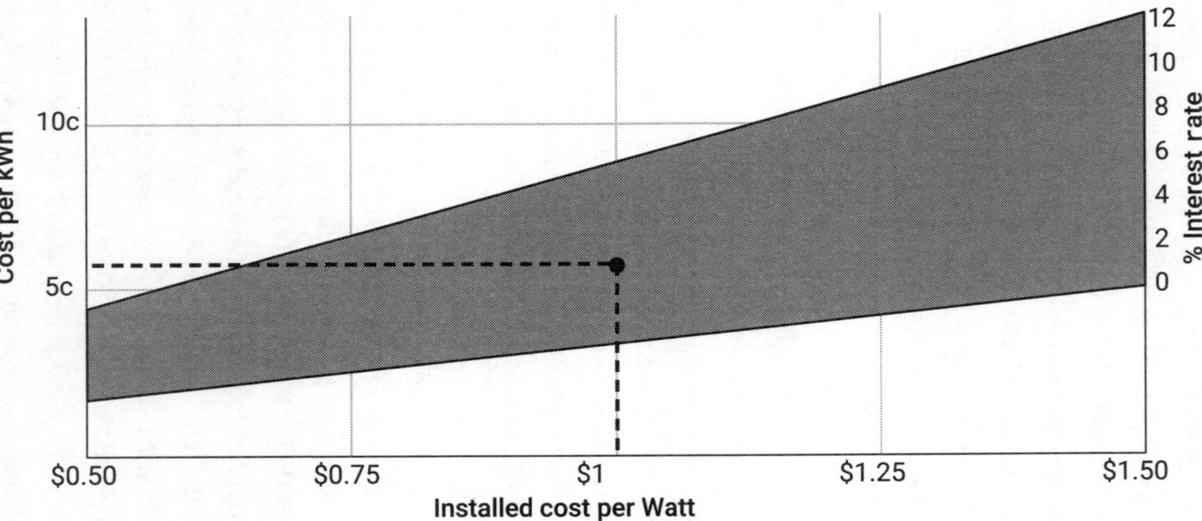

Figure 4.2

The cost of solar. At 6% interest over twenty years, a $1/W solar install produces electricity at about 6 cents per kWh. Lower interest rates and lower costs have a compounding advantage.

only use 50% of your daily solar production, you will effectively be paying double per kilowatt hour. If you export your excess solar to the grid, you will likely get paid a feed-in tariff, but these tariffs are falling and will not always cover the cost of your solar.

In this scenario, you might consider buying a battery or subscribing to a community battery. Even if you're not ready to do that, solar is so cheap now that installing more than you need and "wasting" a little still makes financial sense. Waste is often called "curtailing" in the electricity world, and unfortunately as much as 10% of our renewables is already being curtailed. With clever market reform, whereby neighbours can sell their surplus electricity to each other and EV owners are incentivised to charge their cars during the day, eventually your extra solar won't go to waste, and everyone's costs will come down – which is why market reform is important. We need to level the playing field so that your rooftop solar or the battery in your garage is treated equally in the market, just like a power generation plant – because it is electrically equivalent to one. I'll talk more about this in Part 3 of this book.

Rooftop solar short-circuits (pun intended) the traditional way you buy electricity. Historically, the cost of electricity covered the cost of generation (mostly coal and gas, but increasingly wind and solar), but also the cost of transmission (moving electricity long distances) and distribution (getting the electricity the last few kilometres to your home over the poles and wires), and the costs of the metering and retailer. Solar, however, goes straight from your rooftop to your home appliances and vehicles. As we have seen, on pretty much any financing scheme, solar will come in at a cost one-third or one-quarter of what the grid can deliver.

By way of one more example, we'll look briefly at the financed cost of batteries.

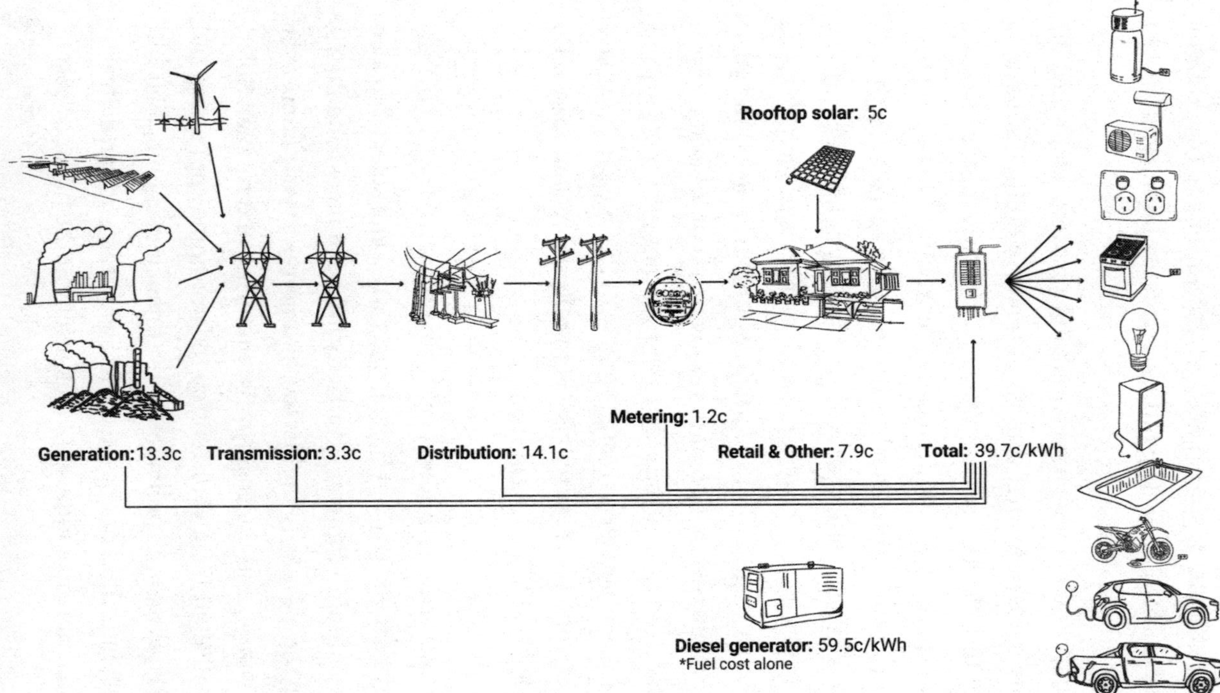

Figure 4.3

The delivered cost of electricity. Grid electricity can be broken down into the component costs, averaging 13.3 c/kWh for generation, 3.3 c/kWh for transmission, 14.1 c/kWh for distribution, 1.2 c/kWh for metering, 7.9 c/kWh for retail, for a total cost of 39.7 c/kWh including the average daily connection fee. The "volume cost" or cost of each extra kWh is around 32 c/kWh. This can be compared to the cost of running a diesel generator at around 60 c/kWh, or the cost of financed rooftop solar at 5 c/kWh. From your switchboard the electricity is routed to your cars and appliances.

Batteries today typically have warranties of ten or fifteen years and generally work reliably for that whole lifetime. We can expect that this will increase over the next decade given the enormous global race to make better batteries. If batteries last twice as long they are effectively half the cost, so this is one of the most important dimensions of battery innovation.

Most manufacturers recommend only using around 80% of your battery to maintain its lifetime. A little bit of energy is lost going in and going out of the battery, as the electricity is converted from AC to DC and back to AC. The remaining electricity is called the round-trip efficiency (RTE) and is usually around 85–90% as the electricity is converted from AC to DC and back to AC. In practice, you are unlikely to use all of your battery's available capacity every day. Sometimes the solar won't generate enough electricity to fill it. The next day you may not use enough electricity to fully discharge the battery. Many people find that they can usefully charge and discharge about 75% of the battery's capacity every day, and that the battery itself loses about 2% of its nominal (or nameplate) capacity every year. We can use this dollar amount to estimate the effective cost of storing your solar energy in a battery.

Figure 4.3 points to when batteries will be cost-effective. The left axis is the cost per kilowatt hour for each unit of energy stored. The bottom axis is the installed cost of the battery. The grey area shows various interest rates applied over a ten-year warranty period. Most batteries currently install at $880–1420 per kWh, but it is widely believed in the industry that batteries will eventually get closer to or even below $250 per kWh – they are already going into electric vehicles at a price below this. The dotted lines show at what installed price batteries will be cost-effective, whether traded on the wholesale market or soaking up your rooftop solar. I have friends, introduced in this book, who are making more than $1000 per year trading wholesale with their battery.

Most likely you are considering a battery because you already have solar. Some people will do it for reasons of independence or resilience, others for notions of ownership, but the question of economics is harder to answer and calculate. If you can generate electricity for 5 cents per kWh, and you can store it for less than 25 cents per kWh, then it is pretty obvious that you can beat an electricity grid that sells to you at 30 cents per kWh (and higher during peak periods). But it will be a few years before the economics are a no-brainer. This highlights the need for the price of batteries and installation to fall, for the lifetime of batteries to increase, and for the utilisation factor to increase. Solar went from a marginal investment to a no-brainer between 2010 and 2020; the same will happen for batteries between 2025 and 2030. A few people who can afford the risk are already making their batteries pay for themselves by letting their batteries play the wholesale market through retailers such as Amber or Reposit.

Acknowledging good policy when it is proposed, we should look at Labor's 2025 pre-election promise to partially fund 1 million home batteries by 2030 with a 30% subsidy as sensible policy. It uses the same mechanism that saw solar go from an expensive luxury to a market-beating commodity. My calculations in Figure 3.2 suggest that the subsidy will pretty much guarantee cost-of-living savings to Australian households who take up the scheme.

Pay cash? Pay fast? Pay slow?

Inherent in each of these decisions is the question of how to pay for it. Broadly, you have three options: pay upfront now; get a short-term loan and pay it off quickly; or get a long-term loan and pay it off slowly. Paying upfront isn't possible for everyone, but it is simple. It maximises your savings because you aren't paying interest, and it means you see

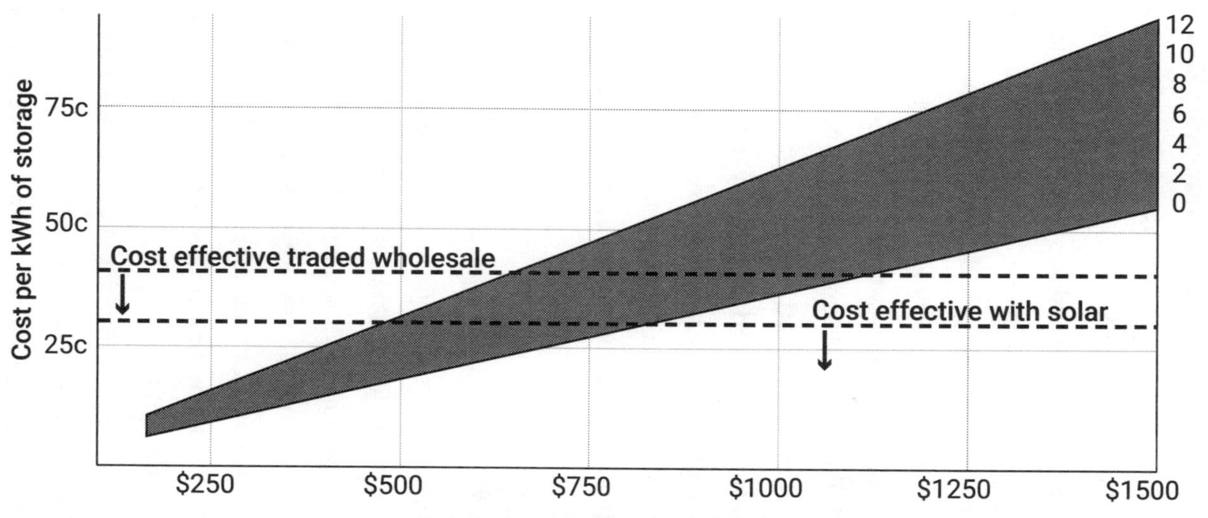

Figure 4.4

The cost of battery storage. Installed cost, interest rate and cycle life determine the per kWh cost of a battery. At ~30 c/kWh, a battery is economic when combined with rooftop solar. At ~40 c/kWh, it can be cost-effective in the wholesale market.

immediate cost-of-living improvements.

If you borrow money to electrify, a loan with a short term and higher repayments will get you out of debt sooner. Once you have paid off your solar panels, the electricity "feels" free. Many people choose a loan that will get them to this free period as quickly as possible without affecting their day-to-day finances too much. This is a matter of carefully choosing the loan period and payment schedule such that the savings generated by electrification are roughly equal to your repayments.

Paying off a loan more slowly makes sense if you know that electrification is a good economic decision in the long run, but you could really use the extra cash in your pocket this week, so would prefer to make smaller repayments over a longer period. You may not save as much money over the fifteen-year life of your water heater or cooktop, but you will still be saving. Many people finance their electrification using their existing home loan. If you have access to this option, it is often the way to get the best interest rate.

Electrifier: Katherine McConnell, finance professional

I'm not a finance guy, I'm an engineer, so I thought it would be wise to talk to an expert in electrification finance, Katherine McConnell of Brighte. Katherine was working at Macquarie Bank when she first saw a solar system, and, while looking at the economics of it, she got excited. She was able to change roles at Macquarie to one focused on energy finance in about 2011. She knew finance was going to be crucial to the future of energy long before I did.

About ten years ago, excited about the potential of solar and batteries, Katherine signed up to be Reposit's first battery customer, purchasing one of the first privately owned batteries on the Australian grid. The battery was an LG, before Tesla had even entered the market. She remembers the first time she turned it on (after a painful installation – no one had done it before!). Her kids ran around the house, turning lights on and off so that they could see which way the electrons were flowing: into or out of the battery, into or out of the grid. The kids thought about it the same way they thought about growing vegetables. Are these veggies from our yard, or from the store?

Although Katherine knew she was an early (perhaps the earliest) adopter, she saw a future where this would be a culturally and financially effective option for everyone. She wrote a blog post about it, and, in her words, "Thank God the blog turned into a business." That business is Brighte, and it specialises in home retrofits and financing, helping people to make their homes cleaner, easier and cheaper to run.

In talking about how people can pay for electrification, Katherine quickly blew my mind. I had always thought of it like a maths nerd or an economist might. What's the interest rate? What's the time period of the finance? What's the payback period? Instead, Katherine emphasised something that now seems so obvious, I'm embarrassed I hadn't thought about it. To real customers, to real people, Katherine assured me, what matters is cashflow. Will this purchase improve my cashflow or make it worse? If I buy solar, will I be able to save enough to buy an extra pizza meal one night a week, so that I don't have to cook?

She describes the energy cashflow calculus like this. Let's say you have a monthly energy bill of $400. If solar can save you $100 a month and it costs you less than $100 a month to own it, you'll end up ahead on cashflow. I do the maths using an online solar calculator like the one at Solar Quotes. Sure enough, it shows that in my postcode, I'll save around $130 a month if I get a 5 kW solar system. I can use a loan calculator to show me that if I pay 6% interest to buy that solar, it'll cost $36 a month if I pay it back over twenty years, or $56 a month over ten years. How you think about it from there onwards is up to you. I could save $44 a month for ten years, then the full $130 a month for the ten years after that; or I could save $64 a month for twenty years. That is about a pizza a week, if only a small one. Alternatively, you could pay the whole thing back in three or four years and then have twenty years of zero-cost electricity. This is the "payback" way of looking at things.

The same maths can be used to figure out the payback period and the weekly cashflow improvements of other zero-carbon solutions – EVs, water heaters, space heaters, batteries, cooktops. EVs, solar panels and water heaters are typically a slam-dunk for cashflow. By the time you've electrified those things, you will only save a little bit more by electrifying the kitchen cooktop and oven. Where these things will save you money, however, is in eliminating the roughly $1 per day required to keep your gas connection. When you kiss your gas connection and its toxic presence in your home goodbye, you will start saving a few hundred dollars each year.

At Rewiring Australia, we have calculated that once a home is all-electric, if about half its energy comes from its own solar panels, the average Australian home probably saves $4000 to $5000 per year.

Katherine has spent some time calculating the payback periods for different home improvements. She estimates three to four years for solar panels. For batteries it depends on a lot of factors, and isn't a clear-cut win yet, as we saw earlier in this chapter. A water heater has a payback period of two to three years. A split system air conditioner for heating that costs $4000–5000 probably pays for itself in five years, but it depends on your home. "If it is a difficult Victorian house with ducted air, it might be twenty years," Katherine says.

The fact is, people's experiences will vary, and Katherine emphasises that different households think about these equations in different ways. Often people nearing retirement choose to pay for their upgrades as quickly as possible, so that when they hit retirement their weekly costs go way down – they pay for their home improvements upfront or choose a short financing period. Younger households sweating from paycheque to paycheque often choose to pay their loans back over a longer period, in order to see an immediate and consistent improvement in their weekly cashflow.

Katherine and I also reflect on the role of a home's "efficiency", which largely means insulation, either batts in the roof, walls and floor, or double-glazed windows. "In certain Australian environments, like Tasmania and Victoria, you need insulation to get the payback. If you don't already have the insulation, it's expensive, and the payback on insulation is likely 100 years." The most cost-effective time to install insulation is when you build the house in the first place or when you are doing a major renovation. This is why we should hold Australian builders to a higher standard – something to which they have historically put up enormous resistance. If we don't build homes right in the first place, it's going to cost all of us a lot.

Katherine might have been an early adopter, but she hasn't yet completed her electrification journey.

She has a home in Sydney and a converted barn for weekend getaways. She drives, and loves, her Tesla Model 3 Performance; to charge it, she just drip-feeds it from a wall plug. She and her kids also enjoy the shared e-bikes you can rent around Sydney. Her hot water and heating are still gas at home and a gas cylinder at the farm; at the farm, she also has a wood-burning fireplace. She has a gas cooktop at home and electric induction at the farm. She just moved into her Sydney home and hasn't yet installed solar there, but at the farm she has a 6.6 kW system and a battery.

5
A Fair Go, Mate? Electrifying Everything for Everyone

Some numbers to consider:

- 31% of Australian households rent
- 16% of Australians live in strata buildings with limited control of their energy assets
- As many as one-third of Australian cars "sleep on the street", meaning they can't easily be charged at home
- Two-thirds of Australian car purchases are used vehicles
- Approximately one-quarter of Australians have credit scores that would make financing anything difficult.

Some people who have picked up this book may be wondering which Australia it was written for. Electrification can sound like a lifestyle choice for the rich, dressed up as climate virtue. Maybe you want a liveable future but can't see how to jump onto the electrification bandwagon. Perhaps you can't get a loan for solar, you can only afford a used car, you don't have off-street parking where you could charge an EV, or you rent or have an obstinate strata committee making everything harder. Then I come along, a guy with enough money to electrify vintage cars as a hobby, preaching about the unattainable.

That is a fair accusation, but a big part of the electrification agenda is to bring everyone along. We need to help disadvantaged communities, low-income households and renters to all enjoy lower bills and cleaner air. We have to save the climate, and we have to do it fairly as well as quickly. Sounds impossible? I believe we can make it inevitable.

I am really proud that Rewiring America helped to design the world's biggest climate plan, the 2022 *Inflation Reduction Act* (IRA) in the USA. But the IRA failed on the equity front, because it provided "carrots" or incentives in the form of tax breaks. This is regressive economics. To qualify for those tax cuts, you have to be earning enough money to be paying the relevant taxes. Many Australian schemes at state and federal level are similar. This adds insult to injury for renters and people who don't own assets. They watch the privileged get more privileged, this time in the name of climate action. If you want to accelerate division and inflame the culture wars, this is a good way to do it.

Similarly, this book is probably going to piss some people off. What doesn't these days? But people have a right to be pissed off. Much of the advice in this book is directed at people who already own assets and have economic power. If you are nineteen years old and passionate about climate change, this book might enrage you, because it recommends a bunch of things, like buying electric water heaters, that you probably won't do until you are at least thirty – and that's only if you are lucky enough to borrow money and buy a house.

If you are a renter, you can have control over which car you own, but you won't have control over most of the rest of the decarbonisation kit because your landlord will. If your landlord makes it hard to install a vehicle charger, even the electric vehicle will be a hard decision to make. So, you can become vegetarian, catch public transport and ride a bike – and those are all good things to do – but you may feel excluded from much of the project to get Australia to zero emissions. And, worse, you won't benefit much from the economic upside outlined in this book.

We really need two policy fronts in Australia: one to address climate change, and one to address our increasing inequality. We can't half-solve climate change, with only rich people able to afford the solutions. In the world we have created, whoever can get credit and finance wins. We need to reinvent economics and the social contract. This means getting into the tax code and eliminating subsidies for fossil fuels and eliminating things like negative gearing – tax breaks for investments that don't produce enough income to cover their cost – which create a divided society of haves and have nots. For climate's sake! We all need a fair go, mate.

If you are a renter, or living in a strata apartment, or facing one of the other obstacles to electrification, there are still some things you can do besides becoming a bike-riding vegetarian. You can buy a relatively inexpensive electric cooktop and cut down on all the wasted energy in a gas cooktop. You can try to negotiate with your landlord, using the arguments in this book, to get him or her to install an electric appliance the next time a gas-powered one breaks. You can convince your parents or other relatives to go electric by telling them why it's so important. You can agitate and organise politically for policies that will electrify everything.

In the early 1990s, gas and electricity bills were only 1.5% of average household spending. Now, it's closer to 3.5%. Back then, when you were deciding to rent a house, the energy bills mattered less than they do today, and there were fewer options to reduce your bills. But a divide has since grown between home owners, who can save money by choosing solar and electrification upgrades, and renters, who are left behind, forced to use the appliances that are cheap for landlords to buy and expensive for renters to run. Unfortunately, everyone is following their economic incentives here, and it will take some pretty bold reforms to change the situation.

Leading up to the 2025 election, the Greens announced a policy that would entitle a renter to compel their landlord to install solar, with the government financing that solar on low, flexible repayment terms favourable to the landlord. The loan would not need to be repaid until the house next sells. This is the kind of policy I'd like to see more of. It would ensure renters get to catch up on the energy upgrades already enjoyed by owners. The renter gets cheaper solar electricity, and the landlord gets the incentive of the value of solar on their property at a favourable interest rate.

Throughout this book, I'll note what options are available to renters now, and what policy changes would make things better. In Part 3, after we've looked at how to electrify the Five Big Things, I'll look at the politics and policy of household electrification: how can we make the energy transition fair and fast for everybody? At the end of the day, the world we live in is the one we design through our tax laws, rules of finance and regulatory environment. To engage in climate or energy work without acknowledging this would be naive.

If you own your own home and are lucky enough to be able to implement the recommendations in this book, I'd also encourage you to work to create a fairer Australia, so that we can all thrive and address climate change together.

Choice

A final point is about "choice". Politicians hate to be perceived as denying people "choice", since it connotes hard-fought individual freedoms. I present the purchasing decisions to electrify everything as choices you could make that will lower your energy bills, but of course these choices also affect our climate. Unfortunately, "choice" makes many people think they should be able to choose the big petrol car or the polluting gas stove, whatever the consequences. Often, people want to make these choices because they are familiar. But these individual choices affect everyone.

In climate terms, if people had no choice but to buy electric appliances when their fossil-fuelled ones died, the world would end up warming around 1.5–2 degrees, which we can perhaps live with. But if we give everyone "choice", and if electrification therefore takes a few more generations, the world will heat up by 2–3 degrees, with much more dire consequences for the earth and for the people and animals that inhabit it. Every fraction of a degree counts.

My point about choice here is that we now have alternatives that enable many of us to make the right climate choices. We need bolder politics and a new engagement with the social contract if we are to choose a better climate outcome, which necessarily will limit our individual choices when it comes to fossil-fuelled machines.

My mate Mike Casey helps farmers to electrify, and he counsels them to "first electrify what you don't love". Many of them have a favourite fossil-fuelled tractor, dirt bike or car that they're not quite ready to part with. By electrifying other things first, they can punt their hardest choice down the road for a while. By the time they get to it, they are more likely to be convinced by economic and other factors that electrifying everything is the way to go.

Part 2

The Five Big Things

Rewiring your life comes down to five big purchasing decisions, each of which comes up no more than once a decade or so. These are the key machines and appliances in your life that use energy and generate carbon emissions. By making smart decisions about these purchases, you'll reduce your emissions, save money and make a major contribution to the clean energy transition.

The Five Big Things are:

- Rooftop solar
- Transportation
- Hot water
- Heating and cooling
- Cooking.

We'll then look at how to make all these electrified components work together.

6
Clean Electricity, Solar and Batteries

Australia truly is the lucky country: the world's cheapest source of electricity is delivered to our homes via rooftop solar. Our abundant solar energy, positive policy decisions and technological improvements have enabled over 4 million Aussie households to install solar, which together accounts for over 11% of our nation's electricity supply.* We can be smug; rooftop solar is two or three times the price in Europe and New Zealand, and four or five times as expensive in the USA.

Figure 6.1 shows just how cheap solar can be, compared with other ways of buying electricity. The day you start powering your home with rooftop solar is the day you will start saving money on your energy bills – and the more solar you can install, the cheaper your all-electric life will be. The same is true for the nation. The more rooftop solar we use in our cars, homes and small businesses, the better for Australia's economy and

---

* Clean Energy Council, "Rooftop Solar Generates Over 10 per cent of Australia's Electricity", 16 April 2024, https://www.cleanenergycouncil.org.au

Figure 6.1

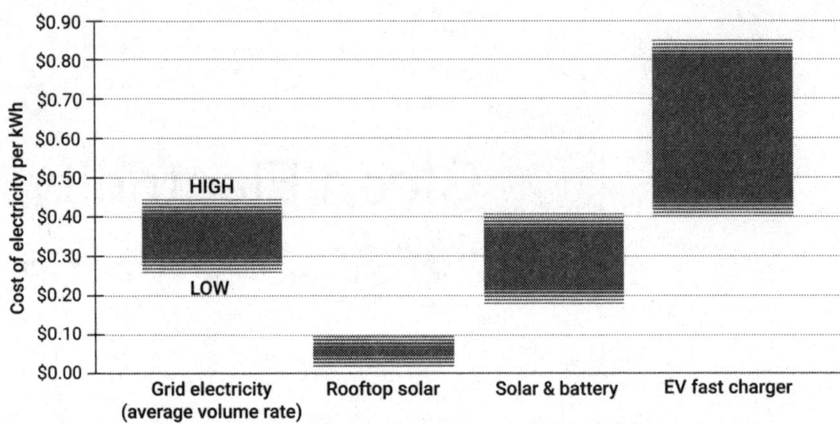

Electricity costs. This shows costs of electricity from different sources, and ranges based on prices in each Australian state and territory.

balance of trade. Increased local power generation reduces the costs of power transmission for everyone, and it is estimated that more than seven full-time jobs are directly supported in Australia for every megawatt of residential solar installed.*

Rooftop solar systems use photovoltaic (PV) panels installed on your roof to convert sunlight into electricity. The cost of solar fell precipitously from 2000 to 2020 and is stabilising now at around $1 per watt installed. That means a 10,000 W (10 kW) system will cost about $10,000.

- **Inverter type**: Your main choice is between microinverters and string inverters. Microinverters attach to the back of each panel.

---

* J. Rutovitz et al., *Renewable Energy Employment in Australia: Methodology*. Prepared for the Clean Energy Council by the Institute for Sustainable Futures, University of Technology Sydney, 2020.

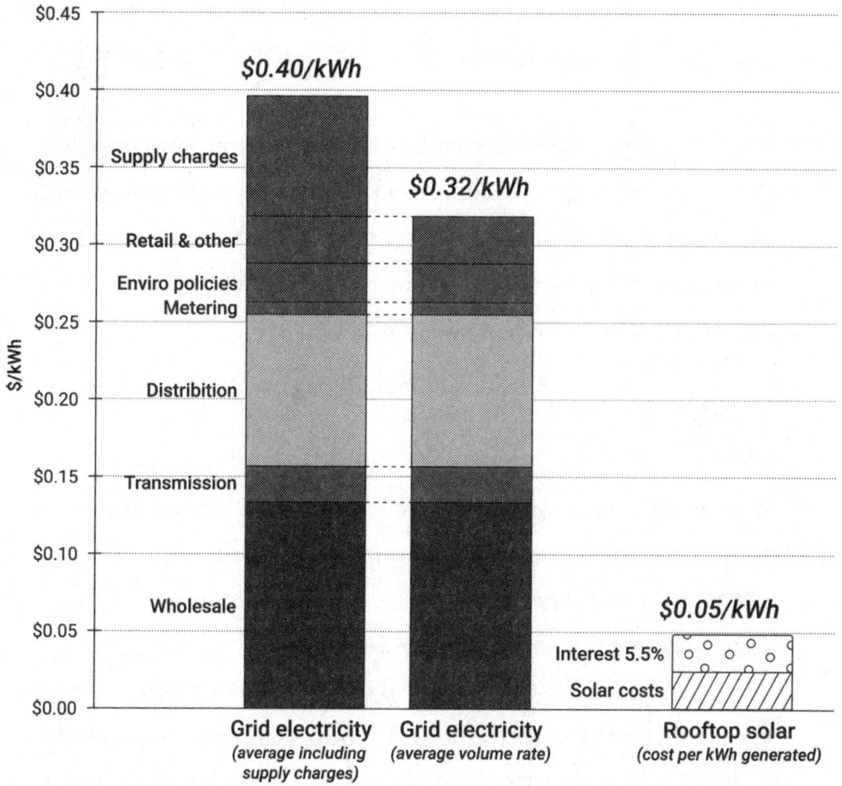

Electricity costs. The cost breakdown of average grid electricity prices in Australia, compared to rooftop solar. Note "supply charges" are normally charged as a daily connection fee, which is why they are excluded from the middle bar. Rooftop solar is only available during daylight hours, so cannot replace all electricity consumption.

With string inverters, each panel is connected to the next, and at the end of the string is a large inverter. Microinverters are generally more expensive but are more effective and get more energy from shady roofs or complex installations. If one module in a string is shaded, it lowers the energy output of all of the panels. With microinverters, each panel produces the maximum it can.

- **Quality**: Invest in high-quality solar panels and components from reputable manufacturers to ensure reliability and longevity. Better quality panels usually have higher efficiency and longer warranties, too.
- **System size**: If you can, choose the biggest system you can afford. This will help you to accommodate new electric needs in the future (e.g. an electric vehicle) and periods of higher energy use, such as winter. It is generally more cost-effective to install a larger system all at once than to add to it later.

Getting solar installed

1. **Assess your solar potential**: A solar provider can do this when they quote, or you can do it yourself using a calculator such as SunSPOT, which will help you to calculate how much solar you can install and the potential costs and savings.
2. **Get quotes**: Request a number of quotes from local providers, or use a site such as Solar Quotes to find recommended installers. During quotation, installers can advise you on which panels and inverters will be best for you based on your roof and location, and they should let you know of any work that may be required beforehand (e.g. waterproofing).
3. **Installation**: Your preferred solar provider will do the installation and connection to your electrical panel and grid.
4. **Monitoring**: Many systems come with monitoring tools to track energy production and energy consumption, to help you to optimise your energy use.

5. **Maintenance**: It's a good idea to have your panels cleaned every couple of years.

Choosing a solar installer

Solar is now big business, so much so that you are probably inundated with more social media ads for it than you would really like. Some people have confided in me that they have lost trust in the industry because the endless offers make it seem like a scam. As I'll reinforce later in this chapter, where I talk to a solar-industry veteran and expert, solar is now so cheap that it is worth paying a tiny bit extra for quality. It reminds me of what an architect once told me about choosing a builder: choose someone you like writing cheques to. Choose someone who has been in the game a while, who gives you good vibes, and who you won't resent paying a small premium for – because a good installer is far more likely to be there for you if there is a warranty problem. If you are nervous about writing the cheques, on the other hand, the operation may turn out to be fly-by-night.

Did you know?

It used to be thought that solar panels were only good for north-facing roofs. With technological improvements and falling costs, these days it can be economical to put solar panels even on south-facing or shady roofs. Your installer can advise you. I placed solar on a vertical wall on my house in California, and the panels produced beautifully, especially in the winter when the sun was low in the sky. Solar is now cheap enough to consider using it like this as an architectural element.

Home batteries

Once you are making all that good solar energy, being able to store it at home will give you the backup you need to rely less on the grid or on fossil fuels – and can even make you money as you sell it back to the grid (and hopefully, in the future, to your neighbours). Battery storage, from household units to utility-scale batteries and electric vehicles, is a game-changer in the energy transition.

Australia's cheap and abundant solar resources have already unlocked millions of savings for households and businesses and reduced our carbon emissions. But to get the full value of our solar resources, we need to capture and store energy to be used when the sun isn't shining. With the cost of batteries falling and their performance rapidly improving, a battery could be the key piece of technology that will enable you to unsubscribe from fossil fuels for good. Whether you're aiming to save money, minimise your carbon footprint or prepare for emergencies, it's a smart time to consider investing in a home battery.

As we saw in previous chapters, at current prices, buying a battery isn't an obvious economic slam-dunk, although this will likely change in future. There are other reasons why people choose to invest in batteries, even if it won't immediately save them money. Maybe having a battery gives you a sense of security and control – you know where your unused solar is going, and you know you have a backup power source if you need it. Or maybe you're keen to expose your battery to the wholesale market; many people report making about $1000 a year this way, selling back to the grid during peak demand periods when the price is high.

If batteries are such a key part of electrification, where should they all go? On the sides of our houses? On the local distribution grid or at the local substation? On local factories and commercial sites? Out near the commercial solar and wind farms?

There are pros and cons to all these sites, and in reality there will be some in all of them. While batteries at substations or on commercial sites can be larger than those on individual homes, households with battery storage are nevertheless becoming a key part of our energy infrastructure. They can lower peak demand, add security during blackouts and reduce the need for expensive large-scale infrastructure.

I asked Katherine McConnell, who you met in the previous chapter, what would make most financial sense – should batteries be privately owned by individuals and located on private homes, or shared? Her answer was interesting. They would likely be cheaper and better maintained if they were on the grid and shared. But, she observed: "I think Australians like ownership ... they like owning their homes, they like owning their cars, and they like being in control."

For now, if you can afford to pay upfront for a home battery, it is probably worth getting one. If you would need to take out a loan, you might want to pause, refer to the chart in the previous chapter, and wait a tiny while until the economics are clearer.

Things to consider when choosing a battery

- **Compatibility**: If you are buying solar panels at the same time, opt for an inverter that is compatible with both systems. If you have an existing solar system, depending on your current inverter, you may need to purchase a new one.
- **Location**: You have to follow Australian standards as to where batteries can be placed. For instance, you can't have them within

60 centimetres of a window or under a floor. Your installer will be aware of these regulations.

- **Climate**: If you live in a cooler climate, you will probably need a larger battery, as your solar production will be lower. If you live in a hot climate, make sure the battery is shaded so as not to overheat.
- **Controllability**: Consider choosing a battery that can be easily controlled by you, or which can be automated to maximise its use.
- **Chemistry**: While lead-acid batteries used to be popular, especially in remote areas, the most widely available home batteries use lithium-ion chemistry (similar to the batteries in laptops and mobiles) because they are more efficient and have higher power relative to their weight/volume. There are subtypes of lithium-ion batteries, and they vary in efficiency and capacity, so it's best to discuss with your installer. Nickel-iron or flow batteries can't compete with lithium-ion batteries on performance or price; while sodium-ion batteries also can't currently but may do so in a few years.
- **Capacity**: To determine what size battery you will need, calculate (or ask your retailer for) your electricity usage and aim for the battery to cover your energy consumption during peak grid times, which is typically 5–9 p.m.

  Also factor in future changes in your electricity usage – for example, if you eventually replace gas appliances with electric ones. One day we will all have big batteries on wheels (electric vehicles) parked in our driveways, but for many this may be a way off yet.

Getting your battery installed

1. **Research**: You may want to use an online calculator such as Solar Quotes or SunSPOT to get a sense of your options and potential costs and savings, tailored to your postcode and energy usage.
2. **Assessment and quotes**: Consult with a qualified solar and battery installer to assess your energy needs and site suitability. The Clean Energy Council oversees the NETCC (New Energy Tech Consumer Code) approval for installers, whose products and services must meet high consumer protection standards. Detailed quotes should outline the components of the recommended system, its cost and the installation process.
3. **Installation**: Installation usually takes one or two days. Your installer will integrate the battery with your existing solar system and electrical infrastructure.
4. **Monitoring**: Your system may have settings you can adjust to maximise your usage, such as setting the battery to charge and discharge at certain times of day. A good monitoring system will assist this decision-making process.

Financing your solar and battery systems

Some ways to pay for your solar installation include:

- Paying upfront. Average systems take between three and five years to cover their installation costs through bill savings, making them a solid investment.
- Talking to your existing lender about adding the installation costs

to your mortgage, or accessing one of their green finance products if they offer any.
- Exploring specialist solar lenders, such as Brighte, Parker Lane or Plenti.

Don't take financial advice from an engineer. Speak to someone who wears shoes and works in finance to discuss your options.

Government assistance

Federal and state governments offer a range of incentives and supports for households and small businesses looking to install solar. These can take the form of grants, interest-free loans or cash rebates. The information below was current at the time of writing. Visit https://www.energy.gov.au for the latest updates.

National: Under the federal government's Household Energy Upgrades Fund, you may be eligible to apply for a loan with discounted finance to afford the installation of appliances.

The national Small-scale Renewable Energy Scheme (SRES) provides discounts on solar installations, although these discounts will already be included in most prices you see advertised, as they are collected by the installer. This program is a "sunsetting" program, which means that the discount offered gets smaller each year, until the program stops at the end of this decade. This program was probably the linchpin in getting Australian solar to be so cheap. I secretly hope we can do something similar to help people install the rest of the electrification kit.

ACT: Under the Sustainable Households Scheme, homeowners, including landlords, are eligible for an interest-free loan of up to $15,000.

NSW: The Peak Demand Reduction Scheme offers rebates of $1600–2500 for household batteries.

TAS: Through the Energy Saver Loan Scheme, interest-free loans are available to help Tasmanian individuals and small businesses access energy efficient products including batteries.

VIC: The Solar Homes Program offers an interest-free loan of up to $8800 for batteries.

Strata properties

Strata owners do have added challenges when installing solar power. However, many of these can be overcome. Some councils have programs in place to support strata buildings to go solar, and the ACT and Victorian governments offer rebates for strata properties. There are some strata-specific resources available online, such as Wattblock.com, to help you get started. There are products available, such as Allume's SolShare, to help split battery usage between individual homes.

Renters

If you are renting, consider discussing solar options with your landlord. Some landlords may be open to installing solar panels, especially as it enhances the property value and could be eligible for a tax deduction. In Victoria at the time of writing, a solar rebate for rental properties program helps to offset the cost for landlords. If solar panels would reduce your electricity bill, you could consider negotiating a small rental increase as a contribution towards the cost of installation.

Unfortunately, asking your landlord for a battery can be more challenging. Our advice is to first pitch having solar panels installed, as the return on investment is just three to five years on average, making it an easier sell. In the meantime, there are portable backup batteries available that may be worth exploring for grid outages.

Alternatively, explore community solar initiatives (e.g. Haystacks Solar Garden) or shared solar programs, whereby renters can subscribe to a solar project and benefit from clean energy without needing to install their own solar panels.

Renters can also purchase 100% green energy through their electricity providers, and some retailers offer cheap daytime energy rates that make use of solar in the grid.

Community and local engagement

Along with installing your own solar, community engagement is vital. Consider joining local solar co-ops or community energy projects. These initiatives not only help individuals to install solar, but also foster a sense of community and shared responsibility for our planet. If, for instance, we put solar panels on top of the schools and parking garages in our community, we'd save enough money to upgrade, say, our recreational facilities. More importantly, all the money we were once sending out of the community by paying for our electricity from the grid and buying petrol at the pump will stay in the community, allowing us to spend more locally at restaurants and other small businesses, revitalising our towns.

My neighbourhood, postcode 2515, is in the process of electrifying 500 homes, to show that it can be done, that it can be done equitably, and that it keeps money in the community that would otherwise be flowing out to the utilities or the petrol pump. For more information about community solar, see https://www.rewiringaustralia.org/community.

And let's not forget the potential jobs in the renewable energy sector. By investing in solar, we're not just powering our homes. We are also creating jobs in installation, manufacturing and maintenance. This is an opportunity to boost our economy while having a positive impact on the environment.

So installing rooftop solar in Australia is more than just a financial decision; it's a crucial step towards a sustainable future. By harnessing our abundant sunshine, we can take control of our energy needs and pave the way for a cleaner, greener Australia. The sun is shining, and the future is bright.

Solar and battery FAQs

How do I make the most of my panels?

- **Monitor**: For the first few weeks after installation, use an app to monitor your energy habits to ensure you are matching consumption to production. You might be surprised!
- **Load shift**: You can maximise your savings by taking simple steps, such as shifting more of your energy consumption to daylight hours. Use the delayed timer feature on appliances such as dishwashers and washing machines, and set timers for hot water systems. If you have an EV, make sure you charge it using your solar to enjoy huge energy bill savings.
- **Get off gas**: Once you have solar, you should remove all gas appliances including your water heater, space heater and cooktop and install efficient, electric appliances. Don't forget to disconnect and stop paying a gas subscription fee.
- **Invest in a battery**: The next step is to consider investing in a household battery.

How do I make the most of my battery?

- Implement smart energy management systems that monitor and control the charging and discharging of your solar battery. These systems can enhance efficiency and prolong battery life.
- To maximise savings, time your battery use so that you are using it at peak times, so that you can avoid buying electricity from the grid when it is most expensive.
- Many batteries work best if only cycled between 80% charge and 20% charge.
- Consider a wholesale market plan that can match your battery usage with grid fluctuations (see Part 3 of this book for more information).

Will I need to upgrade my electricity meter?

A solar system requires a smart meter on your main connection, and if your switchboard is old it may also need to be replaced, but not necessarily. Your installer will tell you if you need upgrades when they are providing a quote.

What about feed-in tariffs?

Any excess solar you're not using is automatically fed back to the grid, unless you have a battery. The feed-in tariff is what your energy retailer is prepared to pay you for this energy. These tariffs have been falling for years as solar becomes more prevalent. Real savings are made if you can use as much solar as possible when it is produced. You can do this by setting timers and using appliances and chargers during the daytime.

Can the grid handle more solar?

Yes. Our energy regulators have not proactively prepared for the rise in solar by encouraging dynamic or smart solar systems to be installed, which has resulted in blunt responses to try to "curb" the rise of solar energy in the grid – adding "remote shutdown" options or, for some suburbs, just blanket-banning new solar connections. However, restricting solar production is not the solution and doesn't recognise the crucial role household solar has played in lowering energy prices for everyone and helping the grid. A holistic approach to energy management is required – one that encourages "dynamic operating envelopes" that give the grid operators a way to fine-tune solar output when required, and more storage in the system. Soon there will be plenty more electric vehicles to charge and more batteries, which will further increase the capacity for more rooftop solar. There is a lot more untapped potential on our rooftops!

Will my solar work in a blackout?

It can, but only if it is set up to do so – and you likely need a battery. By default, solar systems turn off when the grid goes down for safety reasons, otherwise they might be exporting power back onto a grid that is powered down and being worked on, putting the technicians at risk. Some homes may choose to have a solar and battery system that can run independently from the grid when needed via an isolation switch or similar, which prevents energy flowing back to the grid. This type of system provides resilience against natural disasters and long-term power outages. If you have a battery and a couple of EVs, in theory you could operate without the grid for days or weeks, assuming there's enough sun to charge the batteries.

Are home batteries safe?

Yes. Australia has strict standards that apply to batteries, including their location and installation. If these are followed, the risk of fires is very low. Australia now has over 250,000 home batteries installed; there have been just four fires linked to home batteries in New South Wales, one in 2022 and three in 2023. To put that in context, there are about 6500 residential fires in New South Wales every year.

More fire incidents – in the hundreds – were caused by faulty batteries in e-bikes and e-scooters. Make sure you buy these items from reputable brands with high-quality batteries, and make sure you've got a working smoke alarm in your garage or wherever you park your e-bike.

Can batteries be recycled?

Yes! Ninety-five percent of a lithium battery has the potential to be recycled, and second-life battery manufacturing occurs here in Australia.

In the factories in China where most battery cells are produced, only about 90% of the cells pass final performance checks. The fails are swept back to the start of the production line and recycled. So the technology required for recycling is already well established, it's just a matter of scale to make it economical to reprocess Australian cells.

What about going off-grid?

If you want to go completely off-grid, you will need a very large solar and battery system to cover all your energy needs, and you can face challenges such as town planning regulations. There are a lot of benefits for most Australians in remaining connected to a grid system, so we are not advocating going off-grid. In fact, more grid-connected smart homes will make our overall energy system cheaper and more resilient. But having a battery ensures you have more resilience, control and independence while grid-connected.

Electrifier: Bryn Foletta

Bryn Foletta might not immediately strike you as one of Australia's climate heroes. He drives a Chevrolet Silverado 1500 (and not an electric one), and he smiles so much he looks like a guy who just won the trifecta at the TAB.

Bryn and the companies he has built have, however, installed more than 150,000 rooftop solar systems across Australia, spread across nearly every state. His company, Solar 360 Australia, is expanding from solar into batteries, water heaters, vehicle chargers and other components of household electrification. If you want to see him even more animated than he already is (and he always talks as though he has just slammed a Red Bull) ask him about the latest offering available through his national network of solar installers: electric dirt bikes.

I met Bryn at an industry event crawling with salespeople and spruikers. He seemed like someone who partied hard and got shit done. We became fast friends. Bryn has been in the solar industry for over seventeen years and has probably installed more solar than anyone in the Southern Hemisphere. He fell into it when he was becoming qualified to be an electrician and a friend of his dad's told him about solar. He started his own installation team in Victoria just when solar was taking off, and it grew to about twenty-five employees in eighteen months. Now his company has eight stores, and over the years he has trained over 3000 tradies to install panels. His company is a one-stop shop for all things electric, including solar, batteries, heat pump hot water heaters and electric dirt bikes, yet another reason he and I get along.

After first working as an installer for other companies, Bryn got tired of selling sub-par products. “One day, I was standing with an inverter in my hand, just about to mount it, and the customer said, ‘Oh, these are the best inverters, that’s what the salesman told me.’ I knew it was a piece of rubbish, and I thought, ‘I just can’t do this anymore.’” He decided to start selling solar directly to his clients, who at the time were mostly residential. “I lost most of my hair over the next couple of years but managed to create quite a successful little business.”

At the time, the prevailing model was that a retail company would sell the solar system to the client, who would have to find a licensed installer. “As soon as the client had any dramas, it was a blame game – ‘it’s the installer’s fault, no it’s the retailer’s fault’ – and the poor client couldn’t get anywhere.” So Bryn brought the two sides together, creating a retail shop in northern Victoria that both sold and installed solar systems. In 2014, he became an authorised distributor of Solargain, Australia’s largest solar energy specialist.

Bryn’s vision was a bricks-and-mortar store where customers could walk in, look at the showroom and talk to a salesperson. “First we ask the customer what they’re trying to achieve,” he says. Some want backup power with batteries. Others are mainly concerned with lowering their carbon emissions. Some focus mainly on savings, and some on being independent from the grid. “I reckon probably about 40% are just about money, 30% are about the environment, and 30% about backup, making money and making sure they’re all connected.”

Bryn’s team would come up with a plan and then talk the client through all the bits and pieces involved and how to manage them. They could redirect power to the client’s EV, for instance, feed some back to the grid, or install a home energy management system to distribute the energy loads throughout the day.

Most customers, Bryn says, end up saving a lot; the average payback time of a solar installation is roughly four to five years. He predicts that by 2030, the same will be true of home batteries. Rebates on systems vary from state to state, and some offer interest-free government loans – all of which helps bring the upfront cost down. "So on a $10,000 solar system, for example, 10 kilowatts, the customer's getting covered for 60% of it, including that interest-free loan, and only paying 40% upfront," says Bryn (the average solar installation for Australian homes is now just under 10 kW). There are a lot of small fly-by-night companies in the solar world, Bryn says, who likely will be out of business before the warranty expires. "The inverter may be failing, the system's not monitored, or the power bill comes in extremely high, and they find the company has gone into receivership. There's a lot of cowboys in the solar industry." Such companies often don't know how to install solar without damaging the roof, or how to stay compliant with regulations. There's a lot of marketing from these companies, and it's hard for customers to distinguish good from bad.

Prices vary widely too. "It's such a volatile market, we call it the solar-coaster." He recommends all his customers get a few quotes, compare them and then ask the salesmen to spell out the difference. "It's all about educating the customer," says Bryn. "The more they know about the system, the better it's going to work for them too."

Lately, Bryn has had a lot of guys coming into his shop not because they're interested in putting in solar systems, batteries or electrifying their hot water heater, but because he's the only guy around selling electrified dirt bikes by Stark Varg, which Bryn races. "They're obviously more environmentally friendly, but they're also 30% more powerful than the conventional dirt bike – better than the best combustion on the market at the moment. It's the most powerful thing that's ever been produced for its class." The bikes are also a lot smoother than conventional dirt bikes, he says. "Put a professional on them who's been on combustion their whole life, and within minutes they're used to freely jumping a ridiculous amount of length, like 80 feet, 40 feet in the air. They reckon it's a lot easier to control in the air as well."

You don't necessarily picture dirt bike riders as environmentalists. But the power of the electric dirt bike has been convincing some of them to electrify more than their rides. "We've been able to teach them about electricity and charging and where the power comes from, and they're starting to realise, 'Oh shit, I'm getting really high power bills and they could go down a lot if I electrified.'"

I recognise Bryn as one of our tradie climate heroes. He got in there and got (a lot of) jobs done. As to his own personal electrification journey?

Bryn has a 17 kW solar system. Battery-wise, he has 20 kWh of his own iStore product. Bryn "eats his own dogfood", as they say in Silicon Valley – he uses the products he sells. For his water heater he has a 270 L iStore heat pump. He has reverse cycle air conditioning. He now travels so much through regional Australia that he flies a plane he built himself. He and I often discuss electrifying his next plane. He has a Dualtron electric scooter, so he can land his plane and jump on a folding scooter that does 70 km/hr, taking him from whichever airport he is at to his customer's site. He also has three electric dirt bikes. One for him, one for his kid and one for his date or his son's mate, depending on the weekend. Bryn is waiting for an electric pickup truck like the Rivian to enter the Australian market before electrifying his transportation.

7
Electrify Your Transportation

The average Aussie household has 1.8 cars parked in the driveway and spends $5000 each year filling them up. Collectively, our private cars are responsible for 16% of Australia's domestic emissions. They pollute our air; University of Melbourne research found that pollutants from cars cause around 11,000 premature deaths in Australia every year – that's ten times more deaths than from car accidents.* To reduce this environmental impact, your first option might be to walk, ride a bike or use public transport – electric bikes and scooters really hit a sweet spot for short trips. But when you need a car, make it electric.

Figure 7.1 tells the story about driving electrically bluntly – it's cheaper. It's hugely lower-cost per kilometre or kilowatt hour if you charge your electric vehicle from your rooftop solar. Expressed in cents per kilometre driven, it will cost you 20 cents per kilometre to power an average

---

* "Vehicle Emissions May Cause Over 11,000 Deaths a Year, Researchers Say", University of Melbourne, 24 February 2023, https://www.unimelb.edu.au/newsroom/news/2023/february/vehicle-emissions-may-cause-over-11,000-deaths-a-year,-research-shows

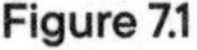

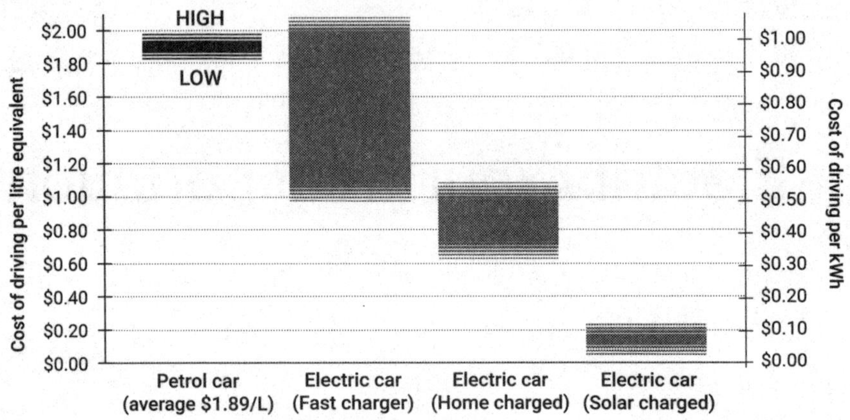

Driving costs. This shows the energy costs of driving. On the left, the first bar shows petrol costs for a petrol engine, then a fast charger into an electric vehicle, a home charger run on grid electricity, and a home charger run on solar.

petrol-fuelled vehicle. To power an EV, you will spend 10 cents per kilometre using a fast charger, 7 cents per kilometre using grid electricity, or between 1 and 2 cents per kilometre using rooftop solar.

We can express this comparison in another way – the cost of fuel per litre. Charging an EV with a fast charger equates to a petrol cost of 80–90 cents per litre. Cruising down the highway powered by electricity from the grid will equate to a mere 40–50 cents per litre. And if you really want to feel smug, an electric bike powered by your rooftop sunshine is equivalent to paying 15–20 cents per litre at the pump. But you will never have to go to the pump again.

My electric vehicle journey(s)

I've now owned seven electric cars and one seriously efficient hybrid. The hybrid was a used 2000 Honda Insight, a tiny, revolutionary two-seater with incredible aerodynamics, an all-aluminium body, and a surprisingly fun five-speed gearbox. It used less than 3 litres of fuel per 100 kilometres. While efficient and fun, it was terrifying to drive on American roads alongside giant F150s and RAM 1500s. We sold it as soon as my wife was pregnant with our first child.

Hybrids were the best alternative available in the early 2000s. Then began the relentless procession of the electrics on our driveway. My first electric was a Fiat 500e. Then I got a Chevrolet Bolt. I converted a 1959 BMW Isetta – a love seat on wheels that I wish I'd never sold. Then I converted a 1957 Fiat Multipla, which I will never sell, because I have learned my lesson.

When we moved back to Australia, we bought the aforementioned strange grey-market import Nissan NEV 200 van. The kids had a love–hate relationship with that one. It was the most amazingly simple and practical vehicle we've ever had, but it must have been a showroom or demonstration model, as it had a picture of an electrical plug about a metre tall emblazoned on its side. Oh, the embarrassment of being driven to school by a father who is an electrification nerd in a nerdy electric van with a giant plug on the side.

In the USA, I had a Rivian R1S "launch edition" – 635 kW of giant American pickup truck. That's over 800 horsepower, which is a different kind of terrifying – far too big and powerful for city streets. This year we bought a far more practical BYD Dolphin for everyday use around Wollongong, where we live.

I remember driving in Larry Page's Tesla Roadster in 2007, which was rattly and behaved poorly; basically, it was a prototype, one of the first dozen they sold. It was a very cool car; its shape was based on the high-performance Lotus Elise sports car. I've driven most of the Tesla models released since and they have become refined and well made. I've driven many of the other electric offerings, even the all-electric Jaguar robotaxis now roaming the streets of San Francisco – a far more competent and pleasurable experience than I'd expected.

I'm a car nerd. My wife laughs at the fact that I can name every European sports car produced between 1950 and 1990 and recognise just about every American muscle car from its shadow. I sometimes wonder at my own hypocrisy. I consider myself an environmentalist, I understand and bemoan the damage that cars (and roads) have done to the world, yet I just can't stop loving the machines. Because of my hopeless attraction to vintage cars, I have arcane skills in fixing carburettors and gapping points. And it's not only cars: I owned a magnificent 1974 Bultaco trials motorbike and just recently bought a 1966 Vespa with a sidecar. (I bought it in Perth, but my wife had me ship it to Wollongong, fearful I would actually ride it across the Nullarbor before I electrified it.)

I hope to do all-electric burnouts at some future Summernats. One has to have life goals. But I also hope that my children will never have to tune a carburettor, replace the oil in an old big-block engine or wrestle with the five fluids required to keep an old Land Rover running. I enjoyed building my first car with my dad, and it was great training in how machines work – and how the world works. I hope I can pass that tradition on to my kids, and I'm encouraging both of them to electrify a vintage car with me to learn hands-on engineering skills. With luck they will grow up talking amps and voltages, motor controllers and MOSFETs. But most people won't build their first electric car. They'll buy it, and that's what this chapter is really about.

Our BYD Dolphin is a very practical little hatchback, and it has become the car that both my partner and I use for anything other than camping or long trips. We've averaged 61 kilometres per day, a little more than the national average of about 40 kilometres. Soon, 85% of those kilometres will be powered by our rooftop solar, costing 1 cent per kilometre. The other 15% will cost about 10 cents per kilometre from fast charging stations. Compare that to 20 cents per kilometre for a typical Australian car running on petrol at its current price. Per day, we spend $1.43 to power our EV. The same daily distance in a petrol-fuelled car would cost us $12.20. Per year, the difference equates to $312 versus $3198.

We have ordered an electric VW bus, which will arrive as this book goes on sale, at which point all of our transport will be electrified.

The economics of electrics

Enough about me and my cars – let's talk about you and your soon-to-be electric car. We all know electric vehicles can be a bit more expensive to buy, but how does that play into the total economics of driving?

I'd like to begin with the observation that driving a new car is not typically an economically rational decision. Cars are an expensive way to get around, and we make them even more expensive than they need to be. If we were all economically rational, we'd all drive Toyota Camrys and Honda Civics, but we don't! Economists like the idea of homo-economicus – that we behave rationally and in our economic self-interest. When it comes to cars, however, we are anything but. Many of us like BMWs and Porsches and Teslas and LandCruisers. The most popular new cars in Australia are the Ford Ranger and Toyota Hilux, neither of which is the cheapest option, and neither of which has an electric equivalent on the market in Australia – yet. So while this chapter breaks down how to make a rational choice about electrifying your ride, we need to recognise that we don't make all our vehicle decisions rationally.

In 2023, of the 1.2 million new cars bought in Australia, 55% were SUVs and 22% were "light commercial" vehicles, which includes popular 4WD utes. For better or for worse, these two categories of larger, less aerodynamic cars dominate our roads. Just 7.2% of new cars bought in Australia were electric in 2023, and 8% were hybrid. Our cars are getting cleaner, but too slowly. Most of the car companies in the world have announced the date by which they will stop making internal combustion engine (ICE) vehicles. Toyota is pretty much the only company that thinks it will still be making ICE vehicles after 2040, and I suspect it will pay dearly for the decision.

Electric vehicles are rapidly improving – they are getting cheaper and they are going farther. In 2011, the longest-range electric vehicles could go about 150 kilometres per charge. The Tesla Model S of 2012 changed everything, setting a new bar with a range of about 400 kilometres. Just a decade later, in 2021, the average range was over 350 kilometres, and the longest-range EVs went over 800 kilometres. That's a huge improvement in just ten years, and you should expect it to happen again. EVs are also getting bigger and taking a range of new forms, such as utes and vans. They are improving in every way. If you aren't seeing what you want yet, wait – but the electric car that does everything you need from at least one of the vehicles in your life probably already exists.

The really expensive thing about all cars is the depreciation. The day you drive your new car off the lot, you lose 10% of the value of the car, which is generally more than you will spend on petrol in the first two to three years of driving it. Each year, the Royal Automotive Club of Queensland (RACQ) publishes estimates of the total cost of owning a host of different vehicles.* In Figure 7.2, I use these RACQ numbers to

---

* RACQ, available at https://www.racq.com.au

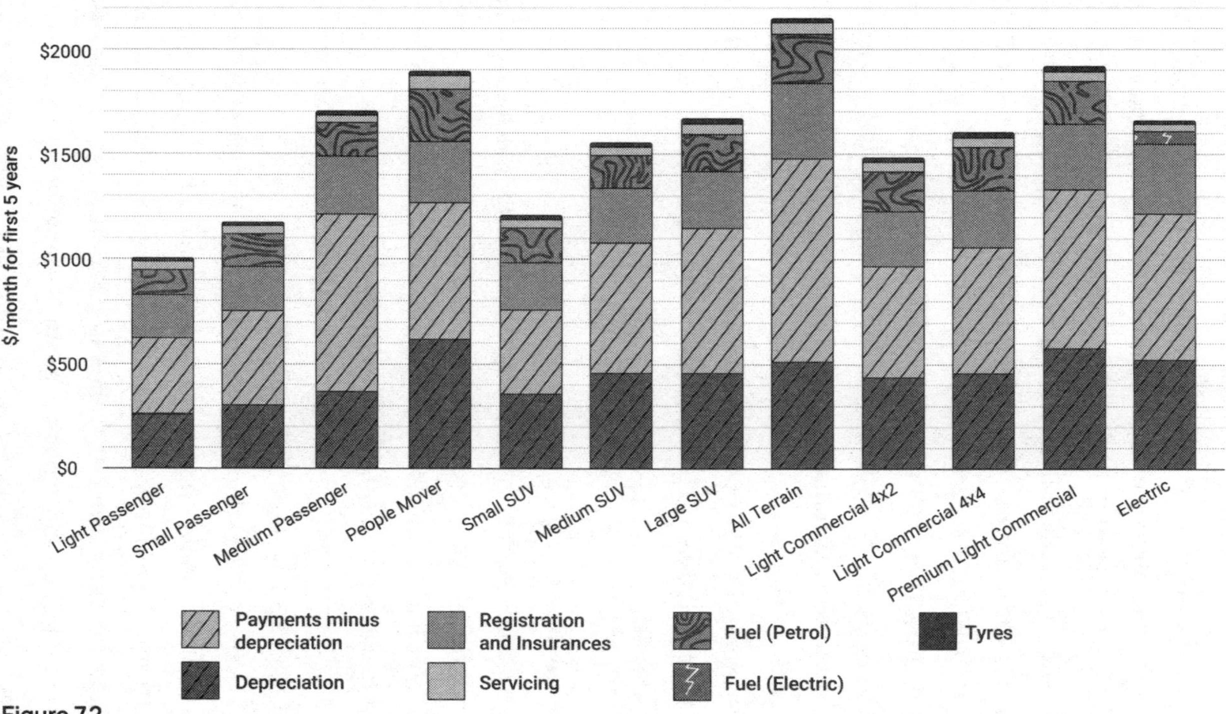

Figure 7.2

Monthly costs of new vehicle ownership over the first five years. RACQ data. Depreciation is the lost value of the car over time. Payments minus depreciation is the sum of the payments made (invested) minus the loss of the depreciation. Registration and insurances are to cover all legal on-road costs. Fuel is either electricity or diesel or petrol. Servicing is the maintenance on the vehicle and tyres are the average replacement costs due to wear.

compare the total five-year cost of ownership for various categories of vehicle, factoring in the cost of payments, depreciation, fuel, rego, insurance, tyres and maintenance. An electric car is obviously cheaper to run than an all-terrain 4WD like a LandCruiser, or a premium light commercial vehicle like a Ford Ranger+, but not as cheap to run as a small passenger car like a Toyota Corolla, or a small SUV like a Mazda CX-5.

The RACQ data averaged the eighteen electric vehicles they looked at, whereas Figure 7.3 looks at the variation between those eighteen electric vehicles. This graph shows that the BYD Dolphin and MG4 are among the cheapest cars in Australia to own and operate. You can likely do even better than this if you run them from your own rooftop solar. If you drive more than average (say 20,000 kilometres per year instead of 13,000 kilometres), your fuel costs will be higher, so you would save even more with an electric vehicle.

Maintenance

Studies now confirm what is intuitively true: electric vehicles are much simpler machines, and consequently the cost of maintenance is lower. An ICE engine has thousands of components and requires extraordinary precision to contain and control a few thousand tiny explosions every minute. An electric motor is just a coil of wire and some electronics, a couple of bearings and an output shaft.

Most electric vehicles don't require a gearbox, because the motor has high torque at all speeds. (Torque is a measurement of your car's force – the more torque, the more power an engine can produce, and the more quickly your car can accelerate from a stop.) An ICE, by contrast, usually has a complicated and expensive gearbox to enable the engine to run at high speeds. Many ICEs have six, seven or even more gears in their

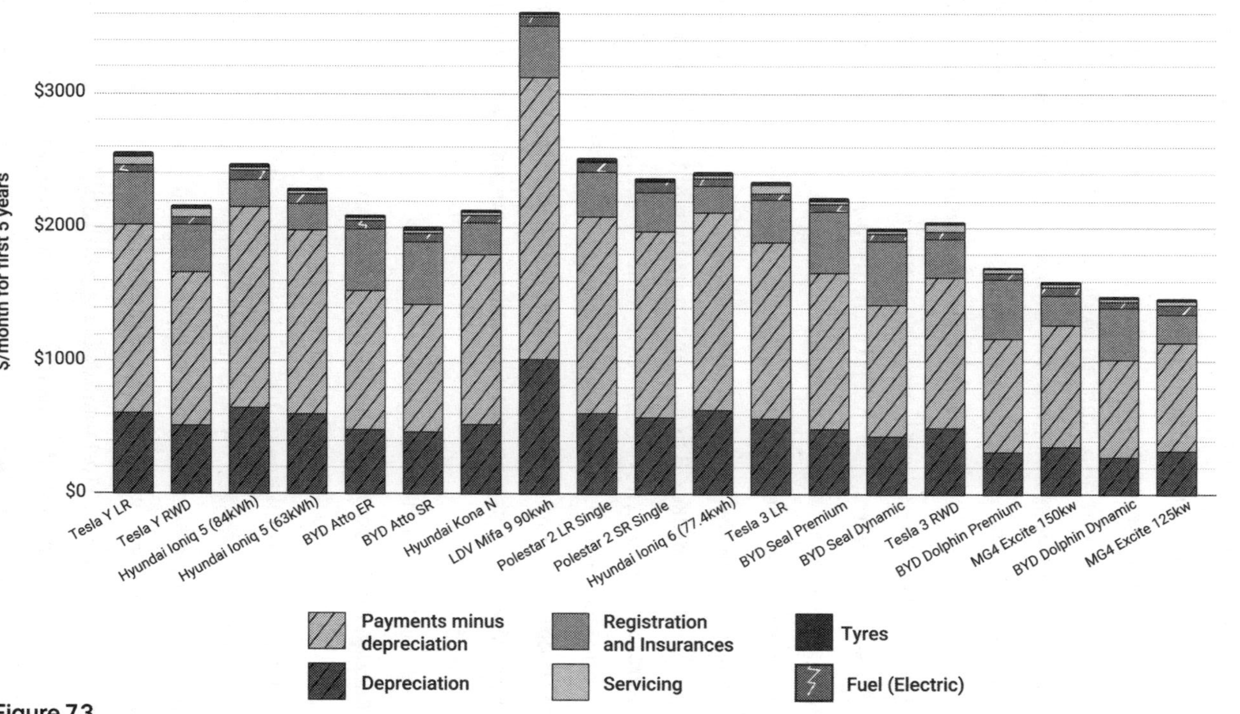

Figure 7.3

Monthly costs of electric vehicle ownership over the first five years. RACQ data. Depreciation is the lost value of the car over time. Payments minus depreciation is the sum of the payments made (invested) minus the loss of the depreciation. Registration and insurances are to cover all legal on-road costs. Fuel is either electricity or diesel or petrol. Servicing is the maintenance on the vehicle and tyres are the average replacement costs due to wear.

increasingly automatic gearboxes. A gearbox costs money, requires lubrication and oil changes, and eventually wears out.

Land Rovers famously leak at least five different fluids: oil, petrol, radiator coolant, transmission fluid, oil in the differentials, lubricating greases and probably wiper fluids too. I've owned a few, and the stains on the garage floor are frequent and colourful. With an EV, there's no transmission fluid, no engine oil, no petrol or diesel, and fewer coolants. Yet again, fewer things to fill up, replenish and replace. My colleague John put carpet down in his garage, so that it can be used as an extra playroom for his kids and as a home office – because he has an EV, there's no dripping oil or petrol to worry about.

As for maintaining the brakes, a lot of the braking in an EV is done by the motor, not the brakes. The motor can be turned into a generator, and the energy that would have been turned into waste heat by brakes is instead turned back into electrical energy in the battery. This contributes to the incredible efficiency of electric driving, and means you will need to replace the brake pads less frequently – another example of lower maintenance costs.

Yes, the battery deteriorates slowly over time and will eventually need replacing, but many EV batteries can now be cycled (charged and recharged) more than 1000 times. If the vehicle has a range of 400 kilometres per charge, this equates to 400,000 kilometres before the battery will need replacing – or longer, if you can cope with a reduced range. Many ICE engines need to be overhauled or rebuilt after shorter distances.

The bottom line is that with an EV, you can expect to save $500 or more each year on maintenance. Most manufacturers will still make you service your EV to maintain its warranty, but you will be less likely to get the dreaded phone call from your mechanic saying they have found something to fix!

Charging at home

The cheapest way to charge your EV is at home, low and slow. We found the factory 240 V type 1 charger to be more than sufficient for our cars. It charges at only 15 amps (15 A x 240 V = 3.6 kW), which means one hour of charging equates to 20–25 kilometres of range. Most cars sit idle for 96% of their lives, so your vehicle probably has plenty of time on its hands to receive the juice.

Type 2 charging

If you want to charge more quickly, you can upgrade to level 2 charging. This is a single-phase, 30-amp, 240 V connection that will give you 40–50 kilometres of range for every hour it is plugged in. A home-style type 2 charger costs around $21,000 to install and won't overwhelm your circuit boards. Increasingly, these chargers are smart and can be programmed to ramp up or ramp down the charging rate, depending on what else is going on in your house.

Three-phase power

For those lucky enough to have three-phase power at home, a three-phase charger delivers 22 kW of charging capacity, giving you 100–120 kilometres of range per hour of charging. If you have a big family or a large home with other large electricity loads, or expect to be disorganised with your charging, it may be worth looking into as an upgrade. If there is already three-phase power on your street, it can likely be brought to your house for $2000–$4000. If not, it could cost as much as $250,000. For more on three-phase power, see Part 3 of this book.

Charging on the road

For most people, charging at home will supply plenty of juice for driving around town. Most charging happens at home – about 85%. Most people drive less than 300 kilometres per week, and EVs now far exceed this range. So, for most people, charging at home, whether once a week or by just plugging in every night when you get home, is enough for day-to-day driving.

For longer trips, you might need fast chargers. In 2022, the publisher of my book *The Big Switch* arranged a book tour in two chunks, with stops on the east and south coasts of Australia. Since I'm an electrification advocate, I didn't want to fly. I wanted to drive an electric car, to see first-hand the extent and limitations of the national charging network. I drove between stops in a Tesla Model 3 and Tesla Model 3 Performance.

I found that it was certainly possible to get where I was going, although at times I had to wait hours while the car recharged. This never really proved to be a problem; you always make stops along the way on a road trip. I travelled with my mum, and recharging gave us a chance to stretch our legs or get a bite to eat. Sometimes, locals stopped and talked with us about our electric car.

Most new EVs have a mapping system onboard that will find the nearest charger and help you plan your trip. There are also apps to help you figure out where to charge on longer trips. The main ones are A Better Route Planner (ABRP), PlugShare, Tesla, Chargefox, Evie, Jolt and Exploren. Enter your destination and they will tell you where to plug in along the way. (PlugShare has a good map of EV chargers across the continent.)

Charging continues to get faster and easier. The standard range Hyundai Ioniq 5 can charge at up to 350 kW so can be topped up from 20% to 80% in around eighteen minutes and have you on your way for another

200 kilometres of highway driving. Most models have a long-range option which can go further and in some cases charge faster.

Rural drivers often have longer commutes, which means electric vehicles are even better suited to their circumstances, because they will save even more on fuel, but they do need faster chargers. Something we've been pushing for at Rewiring New Zealand is installing more fast chargers on farms, with the profit going to farmers. Drivers could stop by a farm, grab some fruit, charge their car in a nice place and then go on their way. This would be a win for everyone. Often farms are close to high-capacity transmission lines, which means minimal upgrades are needed to install fast chargers, compared to urban and town centres, which have larger supply constraints. The Farm Charge scheme would also give farmers an incentive to encourage EV adoption by their friends and neighbours, since every new EV purchased represents a new potential customer. EV chargers could be like gas stations in rural areas – but with all the money going back to locals, who would in turn spend it in their communities and in supporting their farms, rather than to big petrol companies.

Big or small?

Not surprisingly, the first electric cars to compete economically are at the smaller end of the spectrum. There are not many SUVs or 4WDs in the market yet, as manufacturers wait for the prices of the bigger batteries required for 4WDs to come down. I'd like to convince more people to get a small electric hatchback now, and be pleasantly surprised to discover that it more than meets their needs. But if you must have a bigger electric car or a 4WD, they are coming, they are getting cheaper every year, and the choices only keep growing.

Electrifying your vehicles is the single biggest impact your household can make, from both a financial perspective and an environmental one. Electric driving is now the lowest-cost way to drive, including upfront costs, when calculated over the life of the vehicle. The RACQ study only looked ahead five years; if you are going to keep your EV for ten years or more, it is a financial slam dunk, as shown in Figure 7.4.

Driving an EV can save you $1800–$2700 per year in 2025 and $28,000–$45,000 over fifteen years, compared to the costs of running a similar petrol car. The savings on an electric car increase if you can charge it with cheap rooftop solar – but even charging from the grid, an EV easily saves you thousands. An "average" car, according to the Australian Bureau of Statistics (ABS), drives about 11,500 kilometres per year. Our utes and vans drive even more, but are not always considered in average driving numbers, as they are often owned by businesses. If you drive 41 kilometres per day, or 15,000 kilometres per year, charging an electric car with solar is equivalent to paying 13 cents per litre for petrol; charging from the grid is equivalent to paying 80 cents per litre. At the time of writing, petrol prices were around $2 per litre.

EVs are not only cleaner, healthier and cheaper to run. Their ability to work as giant storage batteries will also play an exciting role in our energy future. Win-win-win-win!

The two main types of EV are:

- Battery electric vehicles (BEVs or just EVs), which are pure electric vehicles powered solely by batteries. This is what I'd recommend.
- Plug-in hybrid electric vehicles (PHEVs), which combine a petrol or diesel engine with an electric motor and battery. If choosing a plug-in hybrid, look for one with an extended electric-only driving

range to limit your use of petrol/diesel. Most can go 50–100 kilometres on EV mode which for many people will cover the majority of trips.

There are also hybrid vehicles marketed as "plugless" or "self-charging". These have a smaller battery capacity and rely more on fuel, meaning they will never be a truly zero-emissions vehicle. It is better to choose a battery electric vehicle or a plug-in hybrid.

You may also want to consider how easy (or hard) it might be to re-sell your car in future years. As the upfront price of electric vehicles falls, purchasing a petrol (or even hybrid) vehicle will become less and less rational. Electric cars will continue to be the lowest cost to run, and become more desirable. Petrol cars, on the other hand, will eventually get harder to sell.

Things to consider when choosing an EV

- **Range**: Range anxiety should be a thing of the past. Many people have driven across Australia in an EV without being left stranded! The BYD Dolphin has a range of over 340 kilometres, and the very popular Tesla Model Y long range goes over 500 kilometres. If you don't drive long distances, you may need a smaller range, which means your EV will be cheaper to buy. When setting out on a road trip, there are plenty of apps available to help you plan your charging route. When calculating your range, always remember to take into account factors that may reduce your vehicle's range, such as extra weight if you are towing a trailer or carrying bikes, or extra demand on your battery if you are blasting the heat or air

Figure 7.4

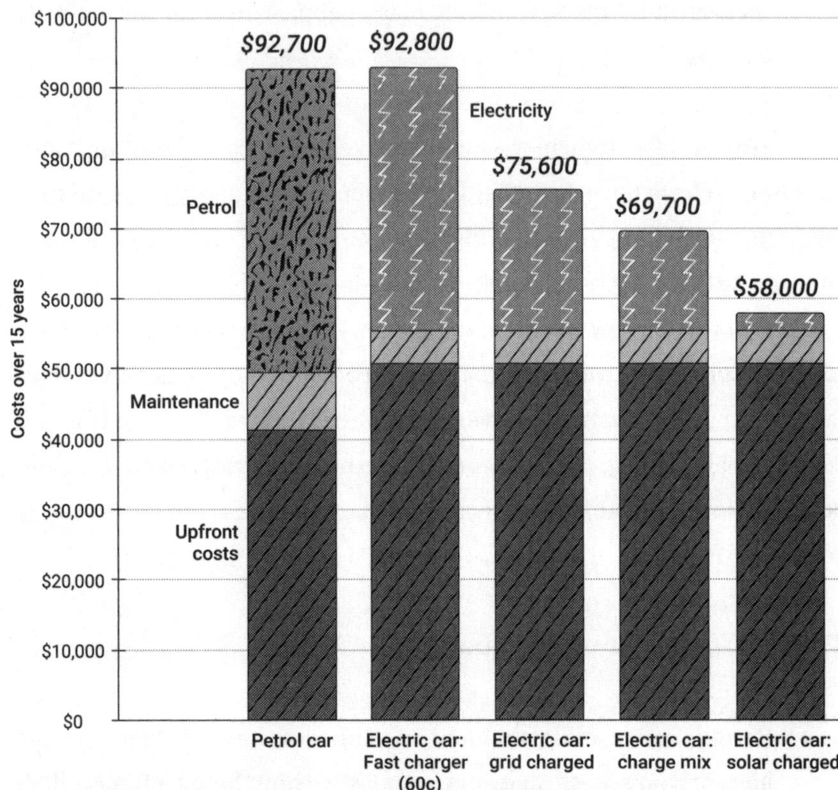

Vehicle costs over a fifteen-year lifetime. The first column shows a petrol car, including the upfront costs, maintenance costs and petrol refilling costs. The next four bars show an electric vehicle, charged using different methods: fast charging only, grid charging from home, mixed charging and solar charging.

conditioning. Driving at high speeds on the highway will also chew up your battery faster.

- **Charging**: You will experience considerable savings if you can charge your car from your rooftop solar during the day. You can charge it slowly using a regular power point, or you can purchase a home charger to charge it more quickly (an average level 2 charger costs about $2000 and will charge most cars in a couple of hours). Most cars only need charging once a week or so, so even if you drive to work you may be able to coordinate weekend or overnight charging or do regular shorter charges. If you don't have solar, time your charging to take advantage of cheaper energy tariffs. Some retailers offer rates specifically for EVs. Even if home charging isn't an option, an EV can still be for you. An increasing number of workplaces and shopping centres offer chargers, and street chargers are becoming more common (lobby your council for more). Fast DC chargers can typically "fill" your car in half an hour or an hour and are good for long-distance driving (few people only rely on fast chargers).

- **Batteries and warranties**: Concerns about EV batteries wearing out and requiring expensive replacements have turned out to be basically a non-issue. Most EV batteries come with a warranty of about eight to ten years, but batteries typically last the life of the car. Depending on factors such as the quality of battery and your charging and driving habits, there can be some degradation. One estimate is that on average, batteries lose about 2.3% of their capacity each year, which means a car with a 340-kilometre range like the BYD might lose around 30 kilometres over five years, yet still go more than 300 kilometres on a charge. You should ask the dealer about the warranties for all aspects of the vehicle you're considering, including basic coverage, battery and roadside assistance.

Financing your EV

One smart way to get a new EV is through a novated lease via your employer (aka "salary sacrifice"). This is because you don't have to pay GST or the Fringe Benefits Tax (if your EV costs under $89,332), which can potentially save you thousands. This option works if your employer is willing to do the paperwork. And it's good to know that because the savings are based on avoiding income tax, you'll save more if you're in a higher tax bracket. As with too many things, our tax system works for people who earn more – but if you are lucky enough to be this person, you could pay forward your good luck to the next generation by getting a zero-emissions EV.

Government assistance

National: Under the federal government's Household Energy Upgrades Fund (HEUF), you may be eligible to apply for a loan with discounted finance to afford the installation of EV chargers. Employers are exempt from paying the Fringe Benefits Tax on EVs.

ACT: Incentives are available for zero- and low-emissions vehicles uptakes, such as discounted registration, exemptions from motor vehicle duty and interest-free loans.

QLD: Rebates up to $6000 are available for zero-emissions vehicles.

TAS: Zero-interest loans are available to install chargers for electric vehicles.

WA: The Zero Emission Vehicle Rebate Scheme offers a $3500 rebate for EVs.

This information was current at the time of writing. Check energy.gov.au for the latest updates.

The forever car

My mate Thor Denmark loves esoteric vintage cars. He has a 1965 VW Beetle that he has owned with his wife for a long time. It's a beautiful cherry colour, with meticulously reupholstered cream leather seats. He and his wife, Hannah, who live in San Francisco, both drive it daily and have for years. Last year, Thor undertook an electric conversion using a kit. Thanks to his typical attention to detail and craftsmanship, he now has a glorious all-electric VW beetle with a battery in its nose and an electric motor in the back, where the engine used to be. The original VW motor had an underwhelming 50 horsepower, and the electrified vehicle has around 250 horsepower, so his Beetle has never been faster – in fact, for the first time, this little old lady can do donuts.

It was a simple conversion, because it was a simple car to begin with. Thor used original VW buttons for the "on", "forward" and "reverse" buttons; the dashboard is now even simpler than it used to be. He upgraded the front brakes to disc brakes. This VW Bug, which did 250,000 kilometres in its first sixty years, should be able to do 1,000,000 kilometres in this next phase of its life. Maintenance is low – periodically reupholster it and paint it, change the tyres when you need to, and you could in principle keep this car running forever.

My forever car is a jaunty little Fiat Multipla 600, which I electrified with six oversized electric skateboard motors. It literally runs on six rubber bands (belt drives) connected to the original drive shaft. If you don't like the increasingly homogenous and vanilla appearance of new cars, with their complicated dashboard interfaces and electronic tracking devices, consider building yourself a forever car and electrifying it. ForEVer?

Electric vehicles FAQs

Should I start with a hybrid?

While hybrid cars may give peace of mind for those who regularly travel long distances, hybrids have limitations compared with full EVs, including more moving parts and maintenance, generally less power and efficiency and, most notably, the car will still be responsible for emitting pollution. Any hybrid car you buy today will likely still be burning petrol in 2040, when we should be aiming to be close to zero emissions. In short, if you can, go straight to an electric vehicle.

The climate would like you to go all-electric. But maybe you want one car that can do all the things your family needs, and one of those needs is long weekends away camping. Or maybe you need to move a large family of six or seven people around frequently. The reality is, the all-electric car that can do those things doesn't exist in the Australian market yet. There are a few models globally that might be able to do what you need, but they're very expensive. In these circumstances, a hybrid is a great idea, but remember there are two types of hybrid, only one of which is efficient and worth buying: the plug-in hybrid.

If you buy a plug-in hybrid in the next few years, I won't judge you. It is a sane choice in a big country if you have a big family and need a big car. A plug-in hybrid means you can still plug the car in and charge it off the grid or off your rooftop solar, and do the first 50 or so kilometres on all-electric.

The other type of hybrid, the "plug-less" or "self-charging" models, uses an electric motor, but the battery is powered by the engine, which burns fossil fuels. Toyota had the early lead in hybrids with the fabulously efficient Prius, but they have lagged in making their hybrids plug-in. I have colleagues working in climate and in vehicle standards who go so far as to describe Toyota's efforts to oversell plug-less hybrids as "evil". I don't think they are evil. But as with everything climate related, I wish everyone would hurry up and focus on deploying the solutions, not the stop-gaps.

Most car trips are very short. American driving habits are pretty similar to ours, and their data tells us that 10% of the kilometres we drive are for trips of less than 6 kilometres. A quarter of our trips are shorter than

13 kilometres, and half are under 30 kilometres. Three-quarters of the trips we drive are less than 50 kilometres. A battery electric vehicle with a range of just 50 kilometres could do most of our trips with zero emissions, and cheaply. That is the promise of a plug-in hybrid. Charge your battery for all the little trips but have petrol in the tank in case the day gets out of hand or the trip is very long and you don't want to stop.

BYD launched a hybrid ute in Australia in 2024: the Shark. The company claims it will drive 100 kilometres in EV mode and more than 800 kilometres on one tank of petrol. There will be a lot of new plug-in hybrids in coming years, but before too long their advantages over electric vehicles will diminish. Batteries will continue to improve, and soon electrics will go the same distance as petrol or hybrid vehicles, but more reliably and at lower cost.

How do I make the most of my EV?

Some useful tips from EV owners include: keep your battery topped up between 20–80% to keep it in its best condition, use regenerative braking (which takes the energy created when the vehicle slows down and uses it to charge the battery), and remember that extra weight, such as bikes on the roof racks, will affect your range, so take that into consideration when planning your charging. Another great feature of most new EVs is vehicle to load, which allows you to use the battery to power regular appliances like a coffee machine or fridge. This is a game-changer for those who want to go camping and not miss out on some of the luxuries from home.

What about the environmental impact of EV batteries?

Mining of any materials should be done as ethically and responsibly as possible. In the past, there were some issues with mining of cobalt for batteries. As the industry has matured, however, so has the traceability and sustainability of mining for critical minerals. Today, more batteries have transparent mineral production or use alternatives such as lithium iron phosphate (LFP) batteries. Australia is in a good position to lead the world in ethically and environmentally produced critical minerals. The best news is that most of the components of an EV battery can be recycled, and the rate of recycling will only improve as the industry expands. In sum, even considering battery production, the environmental impact of EVs is light years ahead of fossil fuel-powered cars.

Should I buy new or used?

We've been talking mostly about new cars in this chapter, but in reality, most of us buy used rather than new. There were around 3 million car sales in Australia in 2024, and only 1 million of those were new cars. Of the used cars sold, only 0.7% were electric. As I write this, there are 195,000 used vehicles for sale on carsales.com.au, only 4151 of which are electric. The average age of a car on the used-car market is about eleven years, so we should start to see more used EVs for sale over time.

The most common EVs on the used market are currently Nissan Leafs and Tesla Model 3s. Early Nissan Leafs came with a 24 kWh battery pack, while later generations had a 40 kWh one. When new, these had a range of around 125 kilometres and 240 kilometres respectively. Because batteries deteriorate, after 100,000 kilometres, you may only get 80–90% of the original range. We had a used Nissan Leaf variation called the NEV200, a jaunty little van with sliding doors. It had a range of about 100 kilometres at local traffic speeds, or 70–80 kilometres at highway speeds. I could drive it from my home north of Wollongong to my parents' house in Sydney (just barely), but I would have to sleep over to allow it to charge overnight. We bought it when it was eight years old, and in the two years we owned it we put 40,000 kilometres on it. It was the most practical car ever, and we rarely had any trouble with the range, because we used our Toyota HiAce for long trips. It was almost free to run, had already done most of its depreciation, and I will always remember it fondly. We used a battery health monitor to test the battery before we sold it. After 100,000 kilometres, it still had 88% of full range.

If you have a predictable commute and daily routine, a used Nissan Leaf with the smaller battery can be had for around $8000 – the cheapest EV you can buy, and very practical. It might be a great second car for a family on a budget, or a first EV for a new driver.

The federal government has tried to incentivise fleet owners to buy EVs, in order to get more used EVs into the market; ex-fleet vehicles are a major source of used vehicles in Australia. They are not really hitting the market yet (it was a 2022 policy), but they should start to do so soon.

On yer (electric) bike

My favourite photo of me is a black and white photo from when I was nine or ten, sitting on a BMX (it was a black and yellow Speedwell), about a metre in the air, having just launched off a hastily built ramp of bricks and an old plank. I'm smiling and concentrating at the same time, and it is obvious from the expression that I think I'm flying a mile high.

That love of bicycles never went away. I've ridden all sorts of bicycles and still enjoy going too fast downhill, braking into corners late and cornering far too close to trees and obstacles. But my love of bicycles has evolved with my life. While it was all about speed and being lightweight when I was young and single, it is all about cargo and carrying things now that I have a family.

Figure 7.5

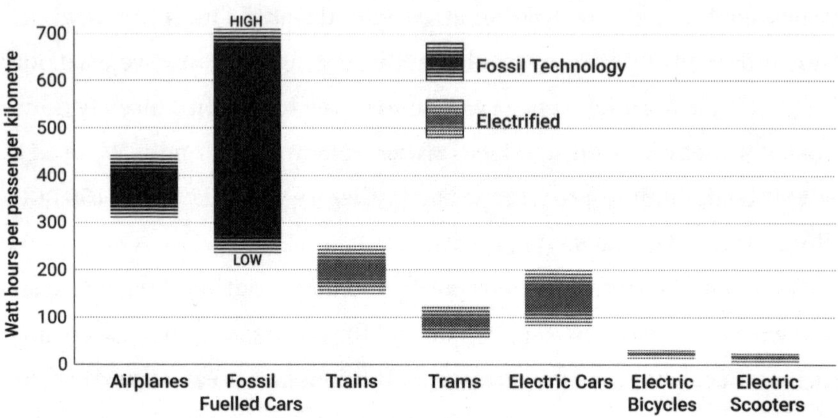

Energy use per kilometre of travel. Includes the efficiency of the transportation combined with the number of passengers.

Which is why I've been ecstatic about the evolution of electrically assisted bicycles, or e-bikes. For most people, most of the time, electric bikes are just better. I find it notable that most of our electrifiers profiled in this book speak enthusiastically about e-bikes above all else, and Fred for example has put 13,500 kilometres on his. I put about 2000 kilometres on my bikes every year.

I have been building my own and buying electric bicycles for twenty years. In that time, they have gone from clunky, niche objects for DIYers and hobbyists to a highly integrated and high-quality product for everyone. I've probably owned fifteen electric bikes, more than ten of which I built myself. I did live in San Francisco where I averaged one stolen bike a year which tells you why I had so many! The good news for you is that if you are e-bike curious, you can now buy, rent or borrow one and skip the building – and hopefully the stolen – part.

I have also been a dedicated bicycle commuter my whole life. When I was young, this meant arriving at work sweaty and either showering or just ignoring my colleagues' disdain for my ripe smell. With the advent of electric bicycles, I get there faster, and the need for a shower is gone, since I'm not working up the same sweat.

If you are fit, young and not carrying very much cargo, I'll be honest: you likely don't need an e-bike, and a pedal-only bike will be faster. If you live low-and-slow, in a place with no hills where you never have to hurry, you don't need an electric bicycle, either. If you want to go fast, carry kids and cargo or just ride comfortably on hilly terrain without sweating, then an e-bike will make your life fabulous. If I am going for a long ride, or going off-road, I will take my road bike or mountain bike. For everything else around town, the electric bike is extraordinary.

Which bike is for me?

There is now a dizzying array of electric bikes out there. If you are after something high-performance, you probably already know what you want, and money may be no object. For everyone else, the key considerations are the ease of mounting the bike, what type of cargo you want to carry, how far you need to go each day, and your budget. I've come to believe the bicycle industry adage that the best bicycle for you is the one you will ride the most. If your intuition tells you a particular bike will get you out of bed in the morning – if it will make you choose to jump on your bike instead of getting into a car – then that is the one for you.

In a practical sense, I think only a tiny number of things determine the practicality of the e-bike you choose. First, of course, you need a bike that is the right size for your body, and anyone in a bike shop can help you to find that. The other important variables, in my opinion, are these ...

Step-over height and step-through frames

The type of frame determines the ease of use of the bike. Step-over frames have a bar across the top. Step-through frames don't have the bar, and you can literally step through the middle of the bike to mount it. Especially for older riders or riders with heavy loads, wrestling with and balancing the bike while swinging your leg over a step-over frame can be difficult. Once upon a time, step-through frames were considered women's bikes, apparently because a woman's skirts could swish without being impeded by a bar. Those days are gone: step-through bikes – or at least the very lowest possible step-over frame you can find – are the best for just about everyone.

Wheels and tyres

The best bike designers I know advise choosing the tyres first and the wheels second and then designing the rest of the bike around them. Those two choices determine everything else about how the bike will ride and the quality of the experience. Tyres and wheels determine how easily the bike will handle different types of surface, its size and weight and how you store it, its ease of use as a commuter bike, and more.

Before good roads, racing bikes had 40–50 mm tubular tyres to cushion the rider from rough roads. As roads got better, the fashion was for thinner tyres. Fortunately, fatter tyres have now come back into fashion; it turns out that 30–50 mm tyres probably roll faster and with less resistance than the skinny ones, which might look fast but are literally going to give you a pain in the arse. More than likely, if you are looking at e-bikes, you are looking for comfort. The sweet spot for tyres is 30–50 mm in diameter. Above about 60 mm, the steering can get a bit "funny", as the big, soft tyre squishes and distorts. Many bike people still think in inches, so ask for a tyre "around 2 inches", or 50 mm. Don't get a knobby tyre unless you are riding dirt tracks all the time; even then, get small knobs. The knobs won't give you extra grip on roads – they'll just be noisy and slow you down. Most people will spend most of their time riding on asphalt and concrete. The tyre should look "slick", like an ordinary car tyre, with most of the texture going around the tyre, not across it.

Bigger wheels can handle bigger bumps. It's fairly simple. At 29 inches, the biggest wheels can basically ride up and down stairs, because compared to the wheel, the stair is small. By contrast, 20-inch wheels don't love bumps. For a long time, 26 inches was common for a mountain bike, or "700 c" (around 27 inches) for road bikes. Don't ask me why we can't all agree on metric – ask the bloody Americans. Anyway, the fashion has been to move to bigger rims, with 27.5-inch rims being a very common, and very comfortable, electric bike standard.

The 20-inch wheel is very strong. That is why they use them on BMX bikes. They can handle a lot of abuse. Many cargo bikes have a 20-inch wheel in the front or the rear. They help keep the cargo weight low to the ground, which improves handling. The other advantage of 20-inch wheels is that they make for a shorter bicycle. Many folding bicycles have 20-inch wheels, and small wheels are a good choice if you are getting on and off trains, putting it in the back of the car, or otherwise have space constraints somewhere in your commute. I love 2-inch tyres on 20-inch wheels.

What about the bikes the kids are riding with 20-inch wheels and 4-inch "balloon" tyres? The bicycle industry, like every industry, has fashions. Fat tyres became a thing in off-road biking and then started appearing on low-cost electric bikes, because a bike with fat tyres often doesn't need any suspension. No suspension means a cheaper bicycle. I wouldn't choose one of these bicycles for myself, because I am a bike snob, but I love how much freedom they are giving people. This appears to be the most popular electric bicycle in my community, and teenagers are using them to carry their surfboards, friends and lovers – sometimes all of those things at once – all over town. What's not to love?

Front cargo, rear cargo, panniers or something else?

I've tried all the cargo bikes. If I was sensible, I would have started a cargo bicycle company twenty years ago and spent my life working on my favourite thing. Then again, maybe it's a good thing I didn't ruin it by turning it into a job. Nevertheless, here are my unsolicited thoughts on cargo options.

You can have your cargo over the front wheel, over the rear wheel or on the sides. For most of the twentieth century, cargo over the front wheel was the classic delivery or "newspaper boy" bicycle. I don't love a lot of weight over the front wheel, but for small loads – a bag of groceries, a six-pack and a bottle of wine, a small school bag or office tote – it can be a fabulous cargo format.

For longer commutes, I really love the cargo over the rear wheel and on the sides, hung low in panniers. I am still using a set of nearly bulletproof Ortlieb panniers that I bought for commuting in 1996. Panniers are bags that go on the racks and mount on the side of the bike – they are indispensable if you are a bike commuter or tourer. This is a great option but limits your cargo to predictable shapes and sizes.

My daily ride is a "bakfiets" style bike which has a long wheelbase and a huge basket in the front. I carry children, dogs, groceries, my lover (my wife), tools and even a shaved ice operation that my daughter runs. I love the big cargo and kids in the front. My wife on the other hand has a longtail bike with cargo in the rear as she finds it easier to handle with heavy loads (like kids). Venus, Mars, front, rear, there is room for both, but you should test them out and see what works better for you.

The kickstand

As you carry more and more cargo, you will care more and more about the kickstand. A kickstand is mostly unnecessary if you don't carry much and are happy to lean your bike against walls and have it fall over occasionally. If you have a dozen eggs in the front basket, however, falling over is the last thing you want it to do. There are one-sided and two-sided kickstands. If you are only carrying light things, such as a dozen eggs, a single-sided kickstand is good enough. If you have a bicycle loaded with children it is beneficial to have a two-sided and very stable kickstand.

Lights

Once you discover the pure joy of an electric bicycle and realise that it is replacing a lot of your car trips, you will likely find that you want to ride it at night. This means you will want lights. You can get cheap battery-powered lights, even rechargeable ones, that you can screw to the handlebar and seat-post. They are good enough. But if you are using your e-bike daily, it is much more convenient if the lights are wired into the bike, so you don't have to remember to charge them separately from the battery. Having all the switches on the handlebar is also very nice.

Brakes

More power means more energy means more speed. More cargo means more weight. More weight and more speed means you need more braking power. You will want good brakes on your electric cargo bike. This means you will want disc brakes (although some very exotic e-bikes have quite capable drum brakes). The bigger the disc brake diameter, the more powerful it will be. On very heavy cargo bikes, you will want 4-piston, not just 2-piston, disc callipers.

Buy good brake pads. You will find that if you are riding an electric cargo bike a lot, especially with heavy loads or on hills, you will go through brake pads very quickly. If you forget to replace them, it's an expensive mistake that will damage the callipers and the disc. Either learn to replace them regularly yourself or schedule regular maintenance.

Electrifier: Dan Bleakley

Dan Bleakley is the host of one of my favourite ever YouTube series. Dan drove his Tesla Model 3 all over Australia, interviewing coal miners and cultural personalities. The result was a hoot – and put to bed the idea that only hippie vegan environmentalists would want to drive electric cars. EVs appeal to pretty much anyone who likes to drive fast.

Dan grew up in Clermont in Central Queensland, one of several coal mining towns in the region known as the Bowen Basin – one of the largest coal deposits in the world. About a third of the community works in the coal mining industry, including several of Dan's friends and family members. Dan spent three months working in a mine while he was getting his engineering degree.

In the early 2000s, he strayed from coal mining. He became interested in renewable energy and electric vehicle technology and joined the climate action group Extinction Rebellion, participating in several protests and even getting arrested for gluing his hand to the office window of Siemens to protest its involvement in constructing a rail line to the Adani coal mine – right near his hometown.

In 2019, Dan bought a black Tesla Model 3 Performance. He wanted to be part of the transition from petrol and diesel cars. In 2021, he drove the Tesla to visit his family in Central Queensland, where his brother Tim was working as a truck driver at Saraji coal mine in Dysart. Tim asked Dan if he could take the Tesla to work for a few days. "I said, 'Sure, but if you let your mates drive it, make sure you film their reactions,'" says Dan.

The next day, Tim sent Dan a short clip of one of his workmates, a dragline operator named Mark, taking the Tesla for a spin. Since mining towns are very small, there are plenty of quiet roads, perfect for test-driving a high-performance EV. As Mark sat in the driver's seat, getting his bearings with the giant touch screen on the dashboard, Tim asked him to bring the car to a full stop – and then instructed Mark to "just plant it". The Model 3 Performance does 0 to 100 in 3.3 seconds, and Mark's head whipped back against the headrest as an enormous, joyful smile spread across his face. That was followed by some light-hearted swearing and surprised endorsements from his three workmates in the back seat.

As soon as Dan saw the clip, he posted it on Twitter and it started going viral. "The next day, I realised that this was a golden opportunity to flip the narrative on the divisive and toxic politics that had been holding Australia back for so long," says Dan. He decided to turn it into a podcast, and *Coal Miners Driving Teslas* was born.

Dan and Tim got to work, taking locals – many of them childhood friends – for their first test-drive in the Tesla and posting the videos online. "We were having an absolute ball, taking our unsuspecting mates out for some silent hooning around town." After ten episodes, a friend suggested he should invite Bob Katter – a politician not known for supporting renewable energy. Dan posted: "Retweet if you'd like me to try to get Bob Katter in the driver's seat. I'll dead-set drive it to Charters

Towers if he says yes." That got around 600 retweets, and a few days later a staffer called to say Bob was keen. Katter was as impressed with the ride as anyone else – his first, second and third reactions were "YEE-HAW".

After the successful drive with Katter, Dan received enough donations to kick off a six-month journey across Australia, producing eighty-seven episodes in five states, interviewing twenty politicians and a few other hilarious guests, such as the former world woodchopping champion David Foster. "Holy shit," Foster said. "It's like a G-force."

Dan said the whole experience instilled in him the importance of conversation. "We need to talk to each other about the kind of future we want to build together." Dan is now building that future himself. So confident is he of the future of batteries in transportation he is building an all-Australian, all-electric, trucking company to clean up our long haul transportation.

8
Electrify Your Hot Water

Water heating is one of the biggest, and most consistent, household energy loads, accounting for on average 24% of a home's energy use, not including vehicles.* This means that electrifying your hot water is one of the most impactful changes you can make, both environmentally and financially. The humble hot water system also brings one of the biggest opportunities in our energy transition. It can reduce gas consumption significantly, and it can act as a "thermal battery", meaning you can often time when to put energy into it. If you have solar panels and an electric hot water system, you can effectively heat your water at minimal cost with zero emissions. Hard to argue with that!

I don't know anyone who has an emotional relationship with their hot water system. We love our cars for reasons of vanity and practicality. We have strong opinions about how we like to cook. We might feel attached to the particular comforts of our preferred home heating system. But

---

* Rewiring Australia, https://www.rewiringaustralia.org/report/castles-and-cars-discussion-paper

when it comes to hot water, all people really care about is that it works, requires little or no maintenance, is cheap and provides water that flows hot and clean.

There are broadly two types of water heaters: in-line and storage. In-line water heaters, sometimes called "on-demand" water heaters, heat the water after you turn on the tap or shower. In-line heaters are typically gas, because you need an intense burst of very high energy to heat the water before it gets to the shower. The gas industry has deceitfully called these "instant" hot water heaters, but they are slower to deliver hot water to the tap than storage water heaters. In-line water heaters have the singular advantage that they are small. About the size of a breadbasket, they can easily fit in a passageway without blocking the passage. Unfortunately this has made them more popular in new building styles that try to maximise the size of the house on the property and build all the way to the edges.

Electric in-line water heaters are a thing, but they require a higher current connection to your mains board, especially to deliver enough

Figure 8.1

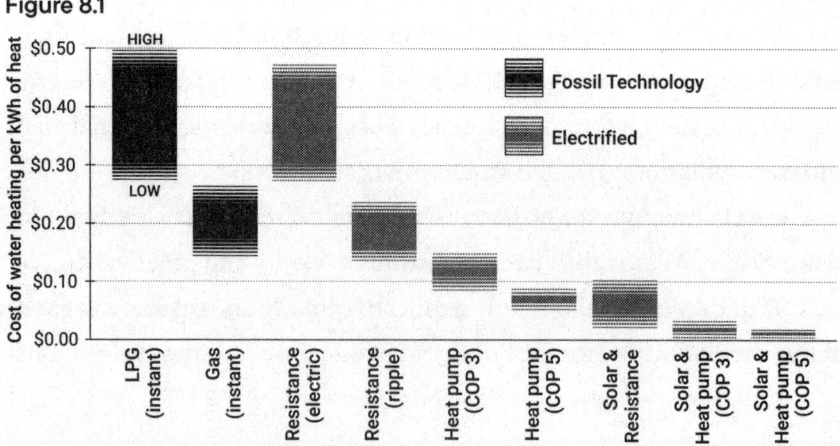

Water heating cost by type of water heater. Includes the cost of the energy and the efficiency of the water heater.

water for a hot shower. To heat a generously flowing shower, flowing at 9 litres of water per minute, from 15°C to 40°C requires about 16 kW. That's a lot. Such bursts of very high power demand aren't good for the electricity system – especially given that most people aren't showering when the sun is high and generating lots of cheap solar.

Storage-based systems are a much better fit for using up cheap energy when it's available. Storage water heaters allow water to be heated any time. The water is stored in a highly insulated cylinder, typically under the house, out the back or tucked away in a closet, laundry or crawl space. Many people will live their whole lives enjoying water from such a device without even seeing it. Some don't even know whether their water heater uses gas or electricity. Most Aussie homes use either an in-line gas system or an electric resistance storage system. However, it is the electric heat pump that will bring the most energy bill savings.

There are four main types of storage hot water heater: natural gas, electric resistance, heat pump and solar-thermal. Let's run through each.

Gas

Gas is easy to obtain but comes with hidden network charges, an uncertain future and higher lifetime operating costs, making it an expensive and risky option. That is why you see a lot more advertisements for gas than for the other options. As your mother – if she's anything like mine – probably told you, "If they're advertising it to you a lot ... you probably don't need it." And because gas hot water systems use a polluting fuel source, they have the highest carbon impact.

Your basic gas hot water heater works very simply. There's a tank with piped-in water. A gas flame heats the water, and when the water gets to a certain temperature, the flame turns down to keep the water warm.

Natural gas is often touted to be "90% efficient", meaning 90% of the gas going into the machine heats the water. However, in real-world testing, their efficiency appears to be a bit lower than this. This impressive number ignores "upstream" inefficiencies such as leaks from pipes, mines and bore holes.

Electric resistance

Electric resistance heaters are close to 100% efficient. They use an electrical element inside the tank to directly heat the water. Sometimes the element – the bit that does the heating – is known as an "immersion heater". Because the element is immersed in water, practically all the energy radiating from it is transferred to the surrounding water. Although some heat is lost while the water sits around waiting to be used, the tanks are very well insulated, which keeps these losses low.

Electric heat pumps

Electric heat pumps are a third, almost magical option. More than 100 years ago, physicists realised you could take energy from one fluid and "pump" it into another. The air outside your home might feel cold to you, but it is warmer than absolute zero, and that heat can be extracted and pushed into your house. Using heat pumps for hot water hasn't been common until recently, but you probably already have two heat pumps in your home. Refrigerators work the same way as a heat pump water heater, but in reverse: the reason there is a puff of warm air at the back of your fridge is because the heat has been extracted from the air inside and expelled. And "reverse cycle" air conditioners are called that because they can either heat or cool air inside your house by extracting or expelling the energy to the outside using a heat pump – it's the same system, run in reverse.

While heat pumps may seem expensive to buy, they are incredibly cheap to run, resulting in savings over the long term. They typically use 65–75% less energy than electric resistance heaters, meaning they are the cheapest hot water system to run. They generally require less maintenance than other systems, with a professional maintenance check required only every 3–4 years – another source of savings.

Heat pumps don't make heat, they move it. They take a small amount of heat from the air outside and pump it into your home. This makes them incredibly efficient. Whereas a resistive hot water system might have 100% efficiency – which sounds pretty good! – heat pumps achieve "efficiency" of anywhere between 350% and 500%, because they are sucking the energy from the air outside to heat the air inside. Thermodynamically speaking, it's not possible to be more than 100% efficient. But practically speaking, a heat pump gives you more heat than the energy you spend harnessing it – so from the viewpoint of your electrical meter, or your bank balance, they are much more than 100% efficient.

Solar-thermal

There are also solar-thermal hot water heaters, which have been hugely popular in Australia. They collect heat directly from the sun using a special module on the roof through which water is pumped, rather than heating the water with gas or electricity at ground level.

These systems are a great way to use the sun efficiently, but they use up the same sunny space on your roof that can host productive solar panels. If you don't have anything on your roof yet, I'd recommend considering solar panels instead – you can use that electrical energy to heat your water, but also to power the rest of your home and an EV. Electrons are more valuable than heat!

What's the best hot water option?

In Australia today, the cheapest and lowest-emissions way to heat water is with a heat pump, using as much solar as you can. In 2025, this set-up would save the average household $400 per year compared to water heating with instant gas. Over fifteen years and with upfront costs included, this would amount to an average saving of $4400.

Figure 8.2

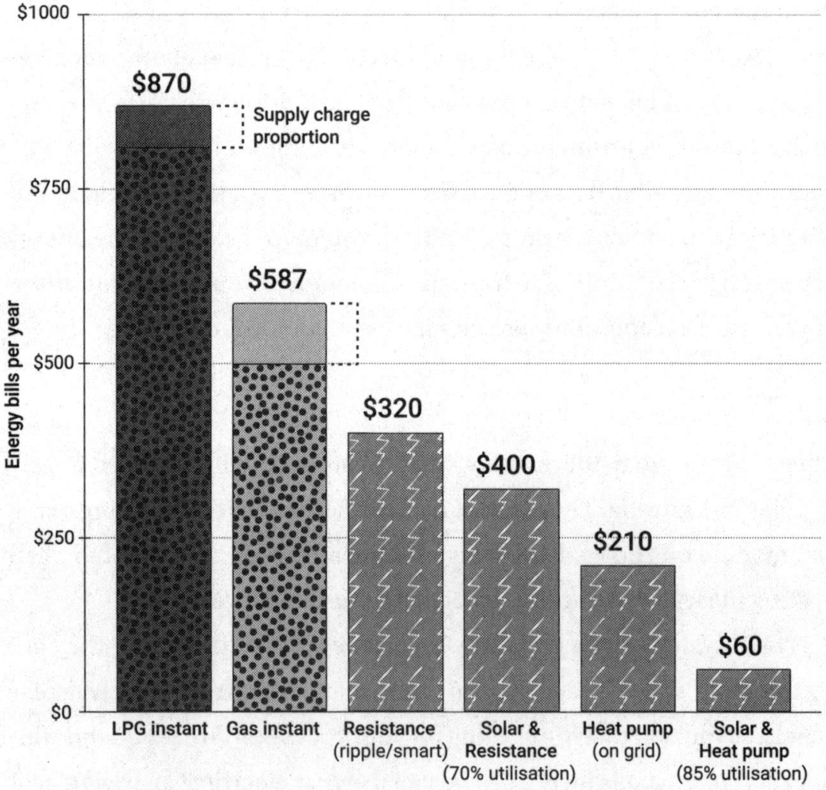

Water heating bills per year. This is the average annual cost to run different types of water heaters. The "supply charge" refers to the fixed daily fees for gas or LPG apportioned to the usage of the fuel of this appliance in the average home.

But if you already have a resistive hot water tank, don't rush out to replace it just yet! If you have solar panels, you can use a timer or smart device to line up your water heating with your solar energy production – and the lifetime savings from this set-up are very similar to a heat pump option.

Figure 8.3

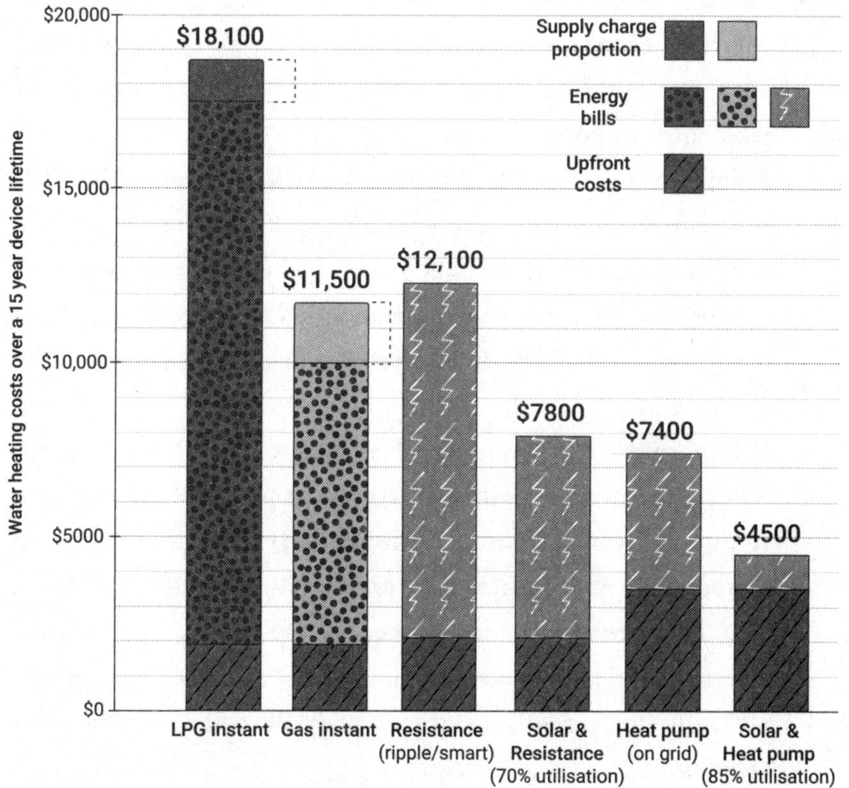

Water heating cost comparison over fifteen years. This is the total cost of water heating with different options over an approximate water heater lifetime of fifteen years. It includes the upfront costs and the operational fuel/electricity costs.

Things to consider when choosing a heat pump

When choosing a heat pump water heater, keep in mind the following factors:

- **Level of control**: A model with a timer or a wired or wi-fi control will enable you to choose when to heat your water, so that it can be powered by your solar panels. This is the optimal set-up from both an emissions and a cost-savings point of view.
- **Design**: Whether you choose a split or all-in-one (integrated) tank may depend on how much room you have. A split system will have a separate, larger compressor, which heats water more quickly but takes up more space.
- **Efficiency**: Look for a high coefficient of performance (COP) rating, indicating greater energy efficiency. The more efficient the system, the lower the running costs. Avoid models that have electric resistive elements built in, sometimes marketed as "back-ups" (there's no need for a backup!), as these models have a much higher energy draw.
- **Refrigerants**: Heat pump water heaters work with refrigerants. The evaporator cell, which absorbs heat from the air, contains refrigerant, which is then pumped through a compressor, which increases the temperature of the refrigerant, and releases that heat into the water. Some refrigerants are greenhouse gases and have a detrimental impact on our climate. CO_2 and R744 refrigerants are good choices for the environment. R410a and R134a should be avoided.
- **Warranty**: Warranties are complex, often covering different periods for labour, tank and parts. In general, opt for longer warranties

and look for a company that has been around for a while. Also note that your distance from approved service agents may add to any repair costs, so find out which agents are approved under your warranty.

- **Noise**: Some heat pumps can be noisy – much like a reverse cycle air conditioner – so consider your neighbours and how close your system may be to any windows. If you are worried about noise, opt for a system with a lower decibel rating (e.g. 37 db).
- **Cost**: Typically, more expensive heat pumps are quieter, use less electricity, have better warranties and use more environmentally friendly refrigerants than their cheaper counterparts.

Getting it installed

1. **Plan early**: Don't wait until your current hot water heater dies. If it's near the end of its life (generally 8–10 years), start researching a heat pump replacement now.
2. **Research**: Check your home's electrical needs. Hot water systems need their own circuit breaker. If you are currently using gas, you may need a wiring upgrade. You can contact a local plumber or electrician for advice. Consider different heat pump models and what will work best for your home. Find out what rebates you may be entitled to.
3. **Get quotes**: Contact licensed plumbers to see if they install heat pumps (not all do – yet). When getting quotes, check that the plumber is eligible for any rebates you may want to claim and that they are able to install your preferred model.
4. **Install**: Lock in your preferred installer. Some plumbers may

subcontract to an electrician. Others will be licensed to complete this work themselves.

Government assistance

National: Small-scale Technology Certificates (STCs) can be generated for heat pumps and solar hot water to reduce the cost of a system. Most installers can claim STCs on your behalf. The more efficient your hot water system, the more STCs you will receive.

ACT: The Energy-Efficient Electric Water Heater Upgrade program offers $500 off the purchase price and $250 credit. The Home Energy Support program provides up to $5000 in rebates for concession-card holders for energy-efficient products, including heat pumps. The Sustainable House Scheme provides no-interest loans.

NSW: The Energy Savings Scheme offers a rebate to upgrade your gas or electric resistive water heater (only via a licensed installer). Up to 100% of the new system may be covered, with a minimum payment of $33.

SA: The Retailer Energy Productivity Scheme provides incentives based on the type of upgrade you choose.

VIC: The Household Energy Upgrades Scheme offers a rebate of up to $1000 via Solar Victoria for households earning under $210,000 a year. The Victorian Energy Upgrades program offers further discounts for heat pumps.

This information was correct at the time of writing. Check energy.gov.au for the latest updates.

Water heating FAQs

How can I get the most out of my heater?

Install a timer or control that enables you to time when you heat your water. If you have solar panels, the best time to heat your water is during the day, when solar production in the grid is at its peak and when the heat pump works most efficiently.

Do heat pumps work in cold climates?

Yes! While heat pumps work more efficiently in warmer climates, they still work effectively in cold climates. The better-quality heat pumps work very efficiently in both cold and warm climates and even when it is snowing outside! Our advice is not to go with the very cheapest versions. Nor must you buy the most expensive – many mid-market options are now excellent.

I've got an electric resistance system. Should I make the switch to a heat pump?

Your electric resistance heater will still use three to four times more energy than a heat pump (which is why we recommend heat pumps). However, if you can power your electric resistance heater from a large solar array during the day, and if your budget is restricted, it may make economic sense for you to keep your electric resistance heater.

What about solar-thermal hot water heaters?

Not to be confused with solar PV panels, which generate electricity, for a long time solar-thermal hot water systems were considered the cheapest form of solar hot water heating in Australia. However, the hugely reduced cost of solar PVs means this is no longer the case, and the solar–thermal system is generally less efficient and takes up crucial space on your roof that could be used for PV panels.

Electrifier: Karl Jensen

It isn't quite accurate to say that no one has an emotional relationship with their hot water system. My friend Karl Jensen spends most of his waking hours thinking about electric hot water systems (and electrification in general).

Karl works at iStore, one of Australia's oldest heat pump providers. He is also a total EV nerd and hands-on engineer, regularly volunteering at electrification and EV information days and posting prolifically on social media, debunking electrification furphies. He runs a podcast called *Just Another Solar Podcast*. He spends his free time drag racing in two souped-up Teslas. To date, his EVs are undefeated.

Currently, Karl says, about 800,000 water heaters are installed in Australia each year, and only 60,000 of them are heat pumps. He would like to see that change, so that most of the water heaters being installed are electric heat pumps.

"Heat pump water heaters are a massive part of our transition to all-electric homes," he says. "It's a previously underestimated sector of the renewable picture." Because they rely on ambient air or ground heat, heat pumps need far less energy, which means far lower running costs and carbon emissions, especially when combined with rooftop solar. They're a key technology in reducing overall energy consumption and transitioning to more sustainable systems in homes and businesses.

Karl recommends a smart hot water heater, which can integrate with a solar inverter or grid. The smart hot water heater makes sure the heater uses the solar when the sun is shining.

Along with his two water heaters and the high-performance Teslas, Karl's electrification journey so far has included an additional three heat pumps to heat his home, an induction cooktop, 26.6 kW of solar from sixty-nine rooftop panels and, of course, an e-bike.

"The e-bike is awesome," he says. Given his love of speed, he wears full body armour and a full-face helmet when he zips around on it. Although he loves his Teslas, "the e-bike is better for your health, for sure".

9

Electrify Your Heating and Cooling

Humans have sat around fires for hundreds of thousands of years. It is hard to change that overnight. But we evolved from wood fires to wood-fired stoves, then to coal, "natural" gas, resistance electricity and, finally, to a highly efficient home heating solution with zero carbon emissions: reverse cycle air conditioning, also known as the heat pump or, sometimes, a "mini split" or split system. (We need a cleverer name for them.)

Space heating and cooling are among the largest energy uses in the average Australian household, accounting for 37% of a home's energy use, excluding vehicles. This varies depending on climate, with colder states such as Tasmania and Victoria spending an even greater share of their energy bills on heating their homes. Alongside water heaters, space heaters are generally the biggest users of gas in a home, making them a key household appliance to electrify for health, financial and environmental reasons. As highlighted by many leading health organisations, gas heaters and stoves emit noxious gases such as carbon monoxide,

Figure 9.1

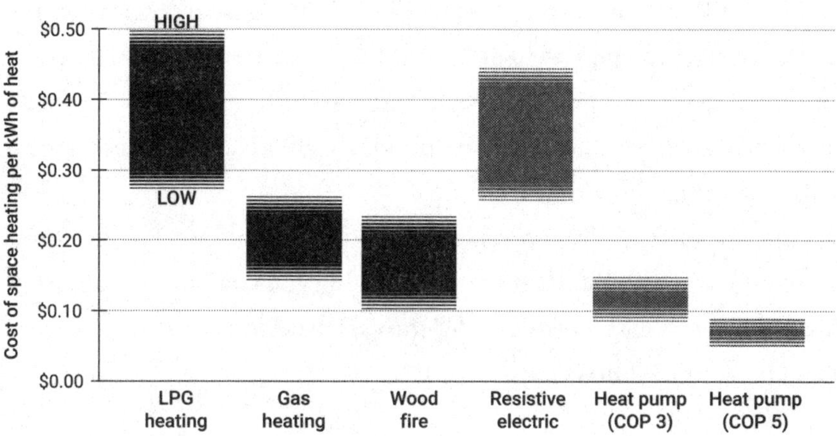

Space heating costs per unit of heat. This combines the cost of energy with the efficiency of each heating appliance.

nitrogen dioxide and formaldehyde. These gases are especially harmful if the heater is unflued.*

Today you have a few options for providing heat. There is good old wood (usually a centralised fireplace in a single room), there is piped natural gas, there is propane, sometimes known as "bottled gas" or "delivered gas", and there is electricity.

There are also two main ways to build heating into your home: centralised forced air, and room by room. In the USA, centralised air is common; in Australia, it is only really common in Victoria and in some mid-twentieth-century homes in New South Wales, South Australia and

---

* For more on the health issues associated with gas, see "Gas Heating: Health and Safety Issues Factsheet", Victorian Department of Health, https://www.betterhealth.vic.gov.au/health/healthyliving/gas-heating-health-and-safety-issues

Tasmania. With central heating, a "boiler" heats a fluid, whether it be air, water or oil, and the heat is pumped around the house in ducts or pipes. The plumbing can be buried under the floor, which means it is architecturally discreet, but makes these homes harder to retrofit, as the ducts are not easily accessible. There are also some efficiency losses associated with moving all those fluids around.

Electric heating, by contrast, gives you more options. Electric solutions are increasingly room-by-room, giving you control over different zones of your house. Let's consider the various electric options, in order from most to least luxurious.

Underfloor heating

The ultimate electric heating solution is underfloor heating. Pipes are laid into a concrete slab or under the floorboards, and the floors are heated from underneath. The temperature required to heat your room is relatively low compared with other heating options, which makes it extraordinarily efficient. The pipes are dead quiet and provide what many people consider the best type of heat, which is radiant, and from below. For people who like to go barefoot around the house (like me!), the floor itself is warm underfoot.

Underfloor heating is quite expensive to retrofit (costing tens of thousands of dollars), but it can be done quite economically at the time of construction. Underfloor heating is most often combined with an air-sourced heat pump.

One nice thing about underfloor heating is that you can use insulated water tanks to store the hot water, this meaning you can heat the water during the day when the sun is shining and use the heat at night. In an optimistic entrepreneurial moment a few years ago, I started a company in this space with some cool technology, but it was "too early". There are now multiple Australian and international companies bringing exactly this solution into the marketplace.

One significant drawback is that underfloor systems are only useful for heating, not cooling. If you were to try to cool your space using underfloor pipes, the cold floor would condense water from the air, making your floors cool, but also slippery and wet.

Air-sourced heat pumps, aka "split systems" or "reverse cycle A/C"

Reverse cycle air conditioning, or "split systems", are very popular and are the most practical and economical solution for most climate zones in Australia. These are heat pumps by a different name. "Split" refers to the fact that the heat pump is installed outside the walls of the house, and the air dispenser itself is on the inside. This way hot air is disposed of outside. These systems are highly efficient in the Australian environment. If you have a large home, it is often sensible to have a few units, to heat and cool different parts of your living space. I'll say more about this in a moment.

Window units

Window units are extremely common throughout the world, especially in Asia and in built-up cities with multi-family housing. The air conditioner sits in a window and draws heat (or cold) from the outside air, through a heat pump, into the home.

Window units are not quite as efficient as split systems, and until recently they meant losing quite a lot of your window. For this reason, I created another company in the USA, Gradient Comfort, to produce units that hang over the window frame. This gives people the efficiency of a split system – and gives them their windows back! This technology has helped to lower heating costs and improve air quality across the USA, particularly in high-density and public housing in New York City. While it's not available in Australia yet, it will be in coming years.

Oil heaters

Nearly everyone can recognise an oil heater; it looks like a bunch of vertical bars sandwiched awkwardly together. Electricity heats up the oil inside, and the device radiates heat from its surface area. Oil heaters might still be your best option for heating an infrequently used spare room, or a cold office space under the house, but they are a very expensive way to heat the whole house, even though the oil heaters themselves are very cheap to buy. Retailers sometimes advertise them as "very efficient", which they are – close to 100%, measured in terms of heat lost – but remember the magic of the heat pump: these can be 300, or 500 or even 700% efficient!

Resistance heaters, bar heaters and panel heaters

There are also the old resistance bar heaters, which radiate a nice infrared heat but smell terrible when some dog hair gets in them. These are less common now, and certainly not the safest or the cheapest option, but their very low upfront cost keeps them in the marketplace.

Modern resistance heaters are often called panel heaters, and heat a plate of glass with electric resistance elements. They are about as efficient as any other resistance heaters, including oil, but the same advice applies as above. They might make sense if you live in a very mild climate where you only want heat a dozen nights a year, or in a spare room that is only used occasionally. If you need heat 50 or 100 days a year or more, it really pays to invest in a more efficient heating system.

Do what's right for you

If I lived in Victoria or Tasmania and was in my home for the long haul, I'd retrofit air-sourced heat pumps, running a hydronic underfloor heating system with storage capacity to absorb my excess solar. But in reality, I live in New South Wales, in an old house that we don't know the long-term future of, so we went with some insulation, curtain and carpet hacks, and a reverse cycle air conditioner. For most Australians, the reverse cycle air conditioner, perhaps with a wood fireplace for the occasional romantic evening, is an excellent solution.

Cooling

In northern Australia, there are very few days when heat is needed, but cooling is essential, and in fact in Queensland homes use far more energy on cooling than on heating. The Northern Territory uses virtually no energy for heating homes, but a lot for cooling. Cooling in southern Australia, particularly Victoria and Tasmania, typically requires significantly less energy than heating. In New South Wales, it can go either way, depending on where you live.

Cooling can already be all-electric with zero emissions. But if you live somewhere that requires both heating and cooling, split systems and "reverse cycle" heat pumps are the way to go. The same efficient appliance can be used to heat and cool your home year-round.

One way to reduce your cooling costs is by using shading, plants, blinds and other passive ways to cool your home. These are incredibly important – but with temperatures set to rise, many of us will also need air conditioners, and that's OK. Our current methods of space cooling are relatively efficient and are already nearly all-electric. We can therefore leave them as they are in an electrified world, as long as we generate that electricity with renewables. The good news is, if it is hot enough to need to run an air conditioner, it is probably also the sunny end of the year. Our existing electric air conditioners will reap the cost benefits of cheaper solar electricity in an electrified home.

Insulation and efficiency

Efficiency has been beaten into us as an unmitigated good. Environmentalists and fiscal conservatives can agree that waste is bad; efficiency is good. But efficiency can be expensive, especially if you are updating an older home rather than building a new one.

Some improvements to efficiency, such as insulation and double glazing, may be prohibitively expensive if you are retrofitting an existing home. Fortunately, there are less expensive changes you can make to create a "tight" or energy-efficient home that doesn't leak heat. A good place to start is sealing all the gaps. Make sure there aren't wide gaps under your front and back doors, seal any cracks in the windows, and fill gaps anywhere there is a connection between your house and the outside air. Your local hardware store will have lots of "strips" and tapes and other handy solutions to keep the cold air out. These are pretty simple DIY projects that you can do yourself, and not impossibly expensive if you hire a tradie. In winter, if you feel a draught, chase it down; fixing it will probably only take a few dollars and a few minutes, and could save you tens or even hundreds of dollars over a few years.

Other efficiency improvements are more expensive, but may still be worth the investment. If the crawl space under your house is ample, getting someone to put in underfloor insulation could cost up to about $5000; if you are handy, you could consider doing it yourself. If you have access to the dead space between your ceiling and your roof, adding more insulation up there is also a pretty cheap job. Putting insulation in your wall cavities is another matter; there are companies that will spray foam in through a small hole in the wall, but wall insulation is done most cost-effectively when the walls are being built or renovated.

You can do a couple of other simple, inexpensive things that will make a big difference to your cooling and heating costs. Planting trees around your house will keep it cooler in the summer and buffer cold winds in the winter.

Shades and curtains are cheap and make a huge difference. In old castles, they often had three layers of curtains: light muslin for summer, a mid-weight cotton and heavy-weight wool or velvet for winter. The layers would trap air between them and be remarkably effective as insulators. My wife and I use heavy curtains to partition the house in both summer and winter. We have a curtain between the side door and the kitchen, creating a small space for shoes and coats and insulating the house from the cold air that comes through our dog door. Another curtain sits between the entryway and the living room.

Cellular or double-walled blinds are a modern miracle and can be used to similar effect. They create insulation in the gap between the two layers, and often have a metallic layer that reflects unwanted sunlight back outside and precious warmth inside. They can be ordered online and are a more affordable alternative to double glazing. This is what we have done for our windows for now. If we decide to knock down and rebuild, that will be the moment to double glaze.

Carpets and rugs can also be remarkably effective. If your house is on the older side like ours, carpet can cover gaps between the floorboards, while providing an extra layer of insulation. Bedouins would layer carpets and rugs in their desert tents to keep them warm in the winter, while also using versions of our curtain trick.

If you have a leaky old Victorian terrace in Melbourne, or a draughty cottage in Hobart, you might use enough winter heat to justify more expensive efficiency measures, but you should look carefully at the maths. Ultimately, it comes down to a balance of cost, comfort and climate. My wife and I have struggled with this; like many Australian houses built last century, our home was built by drunk tradies from chewing gum, bent sticks, asbestos

Figure 9.2

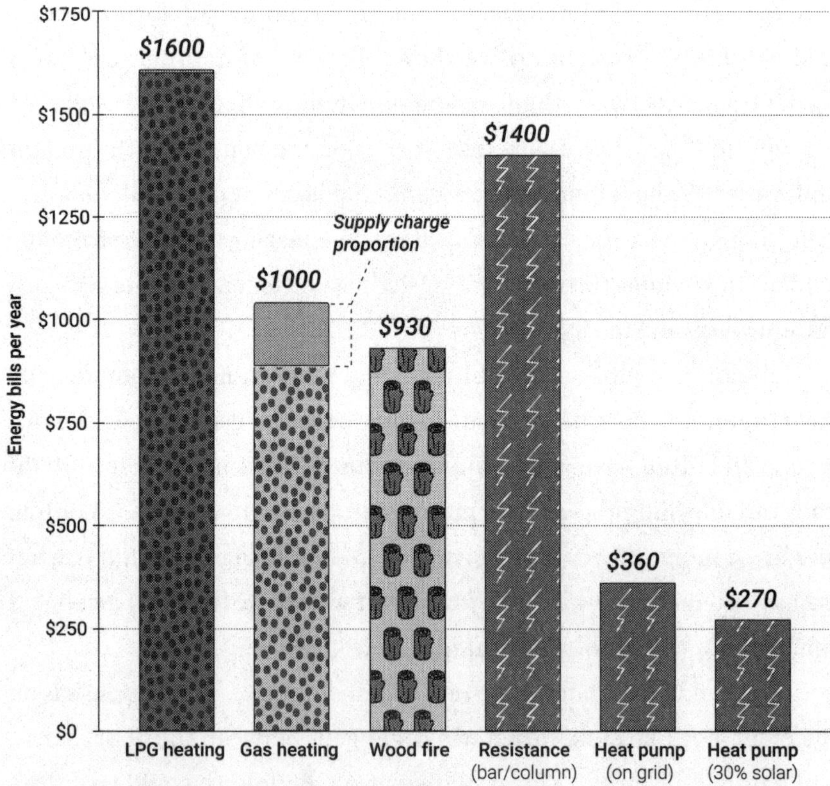

Space heating bills per year. These are the average annual bills to run different types of space heater. The "supply charge" refers to the fixed daily fees for gas or LPG apportioned to the usage of the fuel of this appliance in the average home.

and cigarette butts. It is absolutely charming, and we enjoy living there – but unless we decide to tear it down, adding more solar and another heat pump makes the most sense financially, rather than more extensive and expensive retrofitting. Combined with carpets, curtains, blinds and weather strips this has been enough to produce huge improvements in our energy usage – and the modifications are as charming and eclectic as our old house.

Figure 9.3

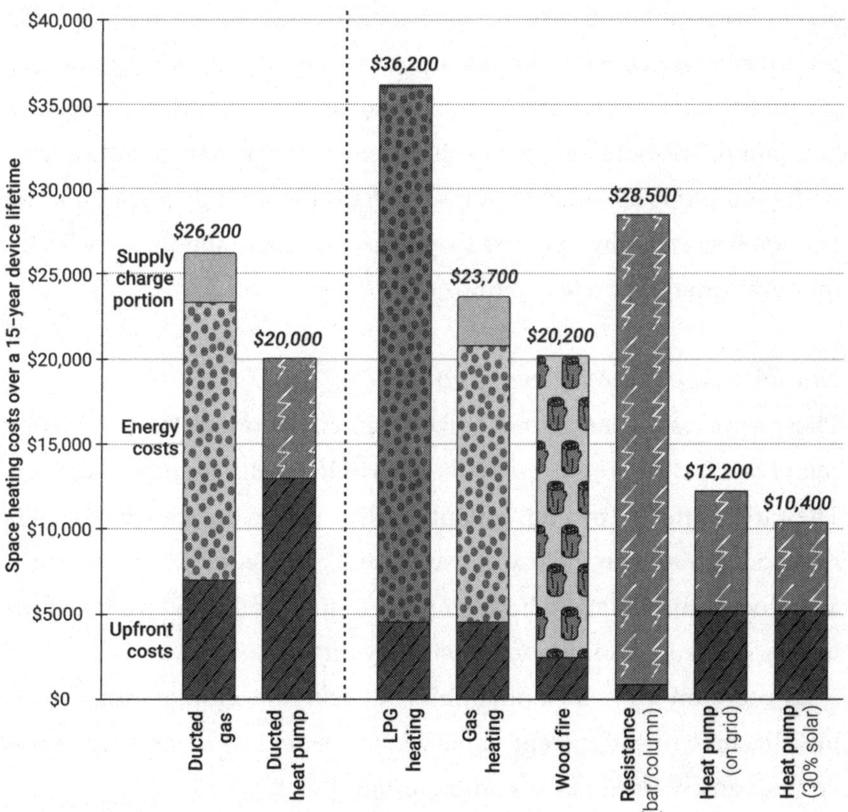

Space heating cost comparison over fifteen years. This is the total cost of space heating with different options over an approximate space heater lifetime of fifteen years. It includes the upfront costs and the operational fuel/electricity costs.

Things to consider: heating and cooling

The main decision you will need to make when choosing a reverse cycle air conditioner is whether to install a ducted central heating system or a ductless split system.

Ducted central heat pumps

We've all seen ducts in buildings: tubes, generally in the ceiling, that are used to move air, whether for heating, cooling or ventilation. Ducted heat pumps use your home's existing ductwork (or new ducts if needed) to disperse heated or cooled air throughout your home. If you already have ducts in your house, this may be the best solution for you, although ducted central heat pumps are more expensive and less efficient to run than split systems.

If your home already has ducts (perhaps for central air conditioning or ducted gas heating), be sure to check with an installer to see whether they are suitable for a heat pump.

Ductless (split system) heat pumps

These are a less expensive but still very efficient option. The term "split" refers to the system's two coils: one goes indoors and the other outdoors. They are connected by a refrigerant line that passes through the wall. The outdoor unit is about the size of a suitcase. The indoor unit is mounted on an indoor wall. You might have multiple units for different rooms, allowing you to create multiple zones within your home.

If you want multiple indoor units but only one outdoor unit, you can install a multi-head system. However, these are less efficient than split systems and require a larger outdoor unit.

Ductless heat pumps are easier to install where there is no existing ductwork. A condenser will be installed outside, and the ductless heat pump heads are usually mounted high on the wall, like old air-conditioner units.

Getting it installed

1. **Research**: Familiarise yourself with the range of options available and identify the key areas in your home that should be targeted for heating and cooling.
2. **Seal leaks**: It's a good idea to identify and seal any gaps where air leaks, and to look at other simple draught-proofing measures.
3. **Get quotes**: Seek quotes from licensed air-conditioning installers. They will be able to recommend the right-sized air conditioner for your needs.
4. **Install and maintain**: After installation, be sure to clean the filter once a year. Blocked filters can reduce the efficiency and operability of your system.

Government assistance

National: Under the federal government's Household Energy Upgrades Fund, you may be eligible to apply for a loan with discounted finance to afford the installation of appliances.

ACT: The Home Energy Support program provides up to $5000 in rebates for concession-card holders for energy-efficient products, including reverse cycle air conditioners and the Sustainable Household Scheme provides no-interest loans.

NSW: The Energy Savings Scheme and Peak Demand Reduction Scheme offer various rebates for eligible households.

QLD: Some incentives are available for south-east and regional Queensland for installing efficient, controllable air conditioners.

SA: The Retailer Energy Productivity Scheme (REPS) offer incentives for installing efficient air conditioners, focusing on low-income homes.

VIC: The Victorian Energy Upgrades program offers rebates up to $2800 to replace inefficient heaters with reverse cycle air conditioners.

This information was current at the time of writing. Check energy.gov.au for the latest updates.

Renters

Renters could choose to purchase a window unit or portable reverse cycle air conditioner themselves, or present the information to their landlord and make a case for installing a reverse cycle system. You could mention any government rebates available, the health impacts of gas appliances, and removing the need to have gas equipment regularly professionally monitored.

To minimise heating and cooling costs, renters can also seal leaks, purchase temporary window glazing, install external shading such as plants, shade cloths or blinds, and close curtains and doors.

Heating and cooling FAQs

How do I get the most out of my heat pump?

- Make sure you clean the filter once a year to ensure it runs effectively and efficiently.
- If you will be using it primarily for heating, you may want to have it mounted on or near the floor. This allows the heated air to flow across the floor, providing immediate warmth.

- If your system has a "dry" or "dehumidify" setting, you can save money by using it to lower the humidity in your home. Unless it is a really hot day, this is often enough to make you feel cooler while using less energy. (For more on this, see the very end of this chapter.)
- Choose a system that has timer functions and/or wi-fi or Bluetooth connectivity to give you flexible control, such as the ability to program the system to run only when the indoor air hits a certain temperature. You may also want to try to time your heating and cooling to coincide with your solar production, to maximise savings.
- Seal any air leaks in your home.

Does a heat pump work in cold climates?

Yes! A heat pump can keep your home warm even when the temperature dips as low as −10°C. In fact, heat pumps heat roughly half the homes in Norway, Finland and Sweden.

Does a heat pump require ductwork?

No. For homeowners with smaller homes, or for those with a need to heat and cool individual spaces within larger homes, mini-split or ductless heat pumps allow you to regulate the temperature in individual rooms. Mini-split systems are perfect for retrofitting homes with non-ducted heating systems.

Electrifier: Sarah Aubrey

Sarah has been an actor and voice actor for decades, who became fascinated with decarbonisation and became an electrification and EV advocate through her social media channel Electrify This! In 2022, Sarah decided to take control of her carbon footprint by electrifying her home, and her life.

Sarah went all-in and did "everything". She took out the gas stove first and replaced it with an induction stove, costing $1700 plus $1000 for the installation. She then took out the fifteen-year-old ducted air; this made room to install underfloor insulation ($3500) and some extra insulation in the ceiling ($2500). She put in four highly efficient reverse cycle air conditioners for about $9500. She installed a tiny solar system of 4.7 kW, and she loves it, load shifting as many things in her house as possible to match daylight hours.

Keen to support local manufacturing and job creation, Sarah bought a heat pump water heater from the Earthworker Cooperative. She paid $6800, but prices have since fallen and there are cheaper options. She loves it anyway, as it is on a timer and is powered by the solar.

Sarah will wax lyrical about the humble ceiling fan. She installed two, and describes them as "so simple and delightful, actually quite magical". Her fans cool her home in summer and "don't even register in terms of electricity use. They're lovely, and silent."

All this electrification enabled Sarah to turn off her gas. I asked her whether she had the pipes removed or simply turned the service off. I admit this was a bit of a loaded question; I was interested in her experience of gas "abolishment", which is an industry term for having the pipes removed from your property. Unfortunately, and no doubt in response to lobbying by the gas industry, the Australian Energy Regulator (AER) has allowed gas companies to charge each Australian household $1256 to abolish. The alternative is paying around $100 for a technician to merely turn off your gas supply. One reason to abolish and remove the pipes altogether is that they frequently leak even if turned off.

Sarah narrated a hilarious outcome. Rather than pay the exorbitant fee to have the pipes removed, she took matters into her own hands. "I had my plumber remove the meter and cap the pipe. I took the meter and drove out to Old Guildford, to the metering place, and plonked my meter on the table. They don't even use computers there, and the guy behind the desk gave me one of those three-layered carbon copies as a receipt for handing in the meter."

She notified her gas retailer that she wanted to leave their service and that she had removed the meter. Somewhat confused, they assumed she must be renovating or moving. A few days later, someone showed up to read the meter and in a panic – when they found it gone – left a note with their mobile number. Sarah sent a photo of the meter with the final reading on it and explained what she had done.

Two weeks later, someone else showed up, scratching their head in her front yard. She showed them the little carbon-copy receipt. Three weeks later, somebody higher up the food chain showed up, looked at her meterless connection, and said, "You can't leave that there, it's not safe." Three weeks later, a crew showed up and removed her connection all the way back to the street. Sarah didn't pay anything. I congratulated her on her civil disobedience. I believe very strongly that Australian households who do the right thing and electrify shouldn't get stiffed with a $1200 bill as they leave the system. I believe this to be uncompetitive behaviour, worthy of a look by the Australian Competition and Consumer Commission, and I believe we should hold the AER to account for giving the gas networks this gift.

Sarah estimates she is saving around $1200 a year by getting rid of gas, and an additional $1000 from her solar. I calculate that all up, she spent about $28,000, including ceiling fans and insulation. Saving $2200 a year (a figure that will increase as energy prices rise) means her payback period is around thirteen years, and that's even though she overpaid for a few of her upgrades.

She is delighted at how quiet and healthy the house is now, which speaks volumes to the advantages of electrification that go beyond just the money; she has also gained comfort and peace of mind.

Sarah rides an e-bike, a Riese & Muller Tinker2. She got it when she sold the household's second petrol car, having realised her household only needs one car. She is planning to replace the other car with an EV this year, but ones challenge will be where to charge it, as she doesn't have a driveway. She lives in Sydney's inner west, and so is reliant on street and public charging, which is neither prolific nor cheap, although it is gradually increasing.

Sarah test-drives EVs for her social media channel, so I was fascinated to know what her pick would be. She told me she'd really like an electrified vintage Mini, but in the meantime she's leaning towards an MG4, because as a rear-wheel drive it handles better than the front-wheel-drive hatches.

Sarah's only bad experience with electrification has been installing solar – a self-described disaster. She only got one quote, which she admits was stupid, and unlike her usual diligent self. (Our solar guy from earlier in the book, Bryn Foletta, insists everyone should get three quotes.) Her first installer put in a tiny 1.7 kW system at high cost that didn't produce much energy, and so she had to have it redone, this time with 4.7 kW and a new inverter.

To get back to heating and cooling – Sarah's best advice is to make sure you understand all the things that these newfangled reverse cycle air conditioners can do. She's a big proponent of "dry" mode. I didn't know it existed until Sarah pointed it out – but your reverse cycle system almost certainly has two modes of cooling. The one you are familiar with cools the air; the other just dries, or dehumidifies, it. The way we experience heat and cold is a combination of the temperature and the humidity. If you dehumidify the air, you will feel cooler. This hack saves huge amounts of energy and money. I should have known my system did this, and once I found out (via Sarah's Instagram account), I was hooked. On all but the hottest days, the dry or dehumidify mode is more than adequate, and far quieter and cheaper.

10
Electrify Your Kitchen

My great-great-grandmother, probably like yours, cooked with a wood-fired oven and stove. My grandmother used to cook with a coal-fired oven and stove. When my mother was a child, she had to collect coal from a pile out the back and bring it into the kitchen; she can still remember the smell and the smoke as the oven heated up. She also remembers when her house became one of the first in the street to install a fancy new gas oven, in the 1950s. My children will remember when their parents and both sets of grandparents replaced their gas ovens and stoves with electric induction cooktops.

Better technologies come along. For a while, everyone is sceptical and thinks the old appliances were better, and a few people hold on to the old technology as though their lives depend on it. But eventually, that old technology will only be found at country fairs – demonstrated once a year, when some oldies or history buffs fire up a steam engine and cook scones over coal, just to show you how it used to be done.

In thinking about electrifying your kitchen, the main thing to consider is your cooktop or stove. I like to cook, and I've come to realise how

Figure 10.1

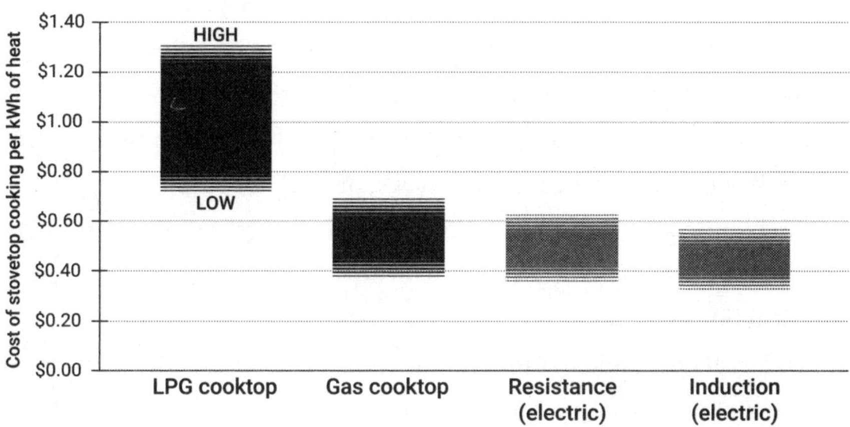

Cooktop operating costs per unit of heat. This combines the cost of energy with the efficiency of each cooktop.

superior an induction cooktop is to a gas one – despite the old propaganda that still makes some people think food tastes better if you're "cooking with gas". The fact is, water boils in about half the time on an induction stove, and the controls make cooking infinitely more precise. No longer will you burn your pots when you forget to turn off the burner; you can set an automatic timer and go read a book. No longer will you leave the house and panic that you may have left the oven on; it will go off automatically. No longer will you have to worry that a child will bump into the hot stove and burn themselves; the stoves aren't hot, just the pots. If your hot chips spill oil on an induction cooktop, it doesn't catch fire. (Yes, I am speaking from personal experience.) There's a reason why more and more professional chefs are shifting to induction cooktops: they're better to cook on.

Induction cooktops work with magnets. You'd have to learn a bit of physics to get into the details, but basically, a coiled copper wire beneath the cooktop's surface carries an alternating electric current, creating an electromagnetic field. This magnetism interacts with the cookware's metal, exciting the ferrous metal, and heating the pan directly without heating up the surrounding cooktop or the air. The cookware heats up from within, which means the cooktop stays cool to the touch. As soon as the cookware is removed from the stove, the heating stops.

Imagine that: you can be boiling water in a pot, place your hand right next to it, and it won't burn you. It won't even be hot. This is the magic of an induction cooktop.

One real downside to induction is that you do have to have the right pots and pans. These are any pans with a ferrous base that a magnet will stick to, which means steel, iron, enamel Dutch ovens, and most other materials besides aluminium and copper. If you have a beautiful copper pot, I'm sorry – you can still hang it up in the kitchen for decoration, or put a plant in it. Woks used to be a bit tricky, but you can find some now with a flat bottom that are induction compatible, as well as countertop electric woks. Induction stoves are far easier to clean than gas stoves. If you're like me, the last thing you or anyone in your family wants to do is spend your time taking the burners off and digging out the grit, grime and crusty bits from underneath. With an induction or electric stove, one swipe over the top and it's clean.

If you're still not convinced that induction is better than the blue flame, consider that gas stoves emit methane, polluting your entire home. This even happens when the stove is off. Invisible methane is constantly spewing into the air you and your children breathe. As the father of two young kids, there's no way I want to expose them to that unnecessary air pollution in their own home.

Stoves are one of the three main appliances that use gas in Australian homes (alongside water heaters and space heaters). While it's the smallest of the three gas users, it has the most direct impact on our health. A 2022 study found that the average gas-burning cooktop emits 0.8–1.3% of the gas it uses as unburnt methane – and more than 75% of this leakage occurs when the stove is off.* The emissions that result from using gas stoves in the USA are estimated to have the same annual climate impact as the tailpipe emissions of 500,000 cars. And it's not just methane. The cocktail of toxicity spewed from a gas stove includes nitrogen dioxide, fine particulate matter, carbon monoxide and benzene, a known carcinogen.

For over five decades, doctors and scientists have been warning about the health consequences of cooking with gas. In the USA, one in eight childhood asthma cases can be attributed to using a gas stove. The fumes from a gas stove put a child at as much risk of asthma as does living with second-hand smoke. If you want you and your children to be healthier, get an induction cooktop! Some induction cooktops may require you to upgrade the electricals in your home. You'll probably need to do this anyway if you are adding other electric features, such as an EV charger. But there are options for people, such as renters, who can't do that. One is to get a single induction burner, which is relatively inexpensive and can be used on any countertop. A friend of mine bought a two-burner induction cooktop and put it on top of her old gas-burning stove. An induction cooktop is the most efficient cooking option and the cheapest, as can be seen in Figure 10.2. But we shouldn't turn our noses up at normal resistive electric cooktops, either. Resistance electric cooktops are healthier

---

* Eric D. Lebel et al., "Methane and NOx Emissions from Natural Gas Stoves, Cooktops, and Ovens in Residential Homes", 56: 4 (2022), pp. 2529–39, DOI: 10.1021/acs.est.1c04707.

Figure 10.2

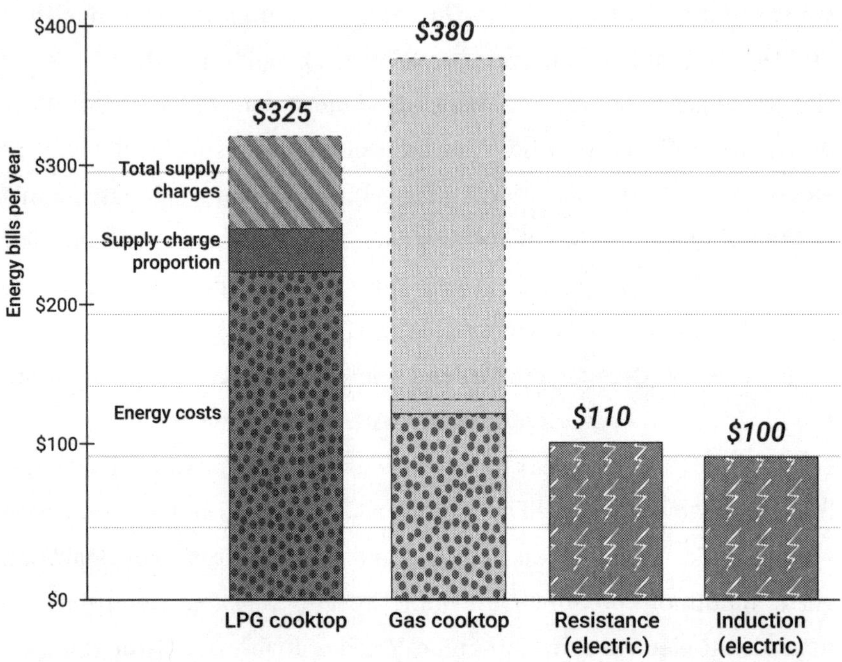

Cooktop energy costs per year. These are the average annual bills with different types of cooktop. The "supply charge" refers to the fixed daily fees for gas or LPG apportioned to the usage of the fuel of this appliance in the average home. The bar above that represents the total gas/LPG connection costs for the entire home. This is relevant, as the cooktop is often the last gas appliance, and disconnecting it can save the home their entire gas connection fees.

and cleaner than gas cooking and have the health benefits of an induction cooktop while being compatible with all types of cookware.

Induction cooktops are far more efficient than both gas stoves and the old electric resistance stoves with coils. Induction cooktops allow 90% of heat to reach the food, as opposed to 65–70% for electric resistive cooktops, and even less for gas cooktops, whose efficiency sits at around 30%. With a gas stove, the flame heats up the ambient air to warm the pan. With an induction stove, only the pan and the food in it get heated.

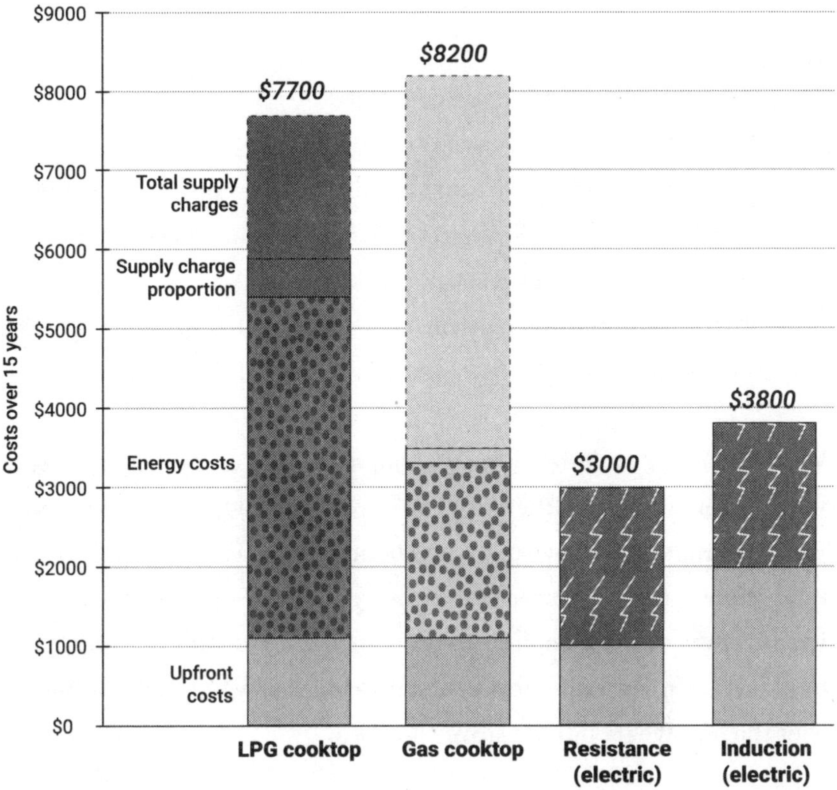

Cooktop cost comparison over fifteen years. This is a cost comparison over an assumed fifteen-year lifespan with different types of cooktop, including energy costs and upfront installation costs. The "supply charge" refers to the fixed daily fees for gas or LPG apportioned to the usage of the fuel of this appliance in the average home. The bar above that represents the total gas/LPG connection costs for the entire home. This is relevant as the cooktop is often the last gas appliance, and disconnecting it can save the home their entire gas connection fees.

Resistance is the middle ground.

Another great thing about electric cooktops is that they can be lightweight and portable. You want one on a table in your backyard to cook for your friends? Just plug it in. And of course, you can also get an electric barbecue. That's next on my wish-list.

Things to consider when choosing a cooktop

Type

There are two types of induction cooktops:

- Portable cooktops that don't require installation and can be plugged into a power point and sit on a benchtop. These are cheaper and, as the name suggests, portable.
- Installed cooktops, which need to be installed by a qualified electrician.

When choosing an induction cooktop, try to find one the same size as your current cooktop, otherwise you may need to adjust your benchtop. Be aware that if you have cut stone benchtops, it is now difficult if not impossible to cut a new space for a stove, as we have discovered that engineered stone creates respiratory problems. In our house, we solved this by placing our new induction stove into a block of wood, and placing this over the hole where our old stove had been.

Cost

There are induction and resistance cooktops at almost every price point now. My colleague Josh Ellison has researched the price points of dozens, and has shown that you could spend as little as $500 or as much as $5000 or even $10,000 – a similar range as for gas. All of them can cook your dinner.

As with cars, our decisions about cooking aren't always purely rational. If cost of living were your only consideration, you would buy a cheap induction or resistance cooktop, because they are two or three times more efficient than gas and enable you to disconnect from the gas network. But rest assured – if you want a very expensive, very high-performance induction cooktop that you can boast about the next time you are entertaining Macquarie Bank executives, there is one in your price range that I guarantee will be better than their high-end gas stoves.

An electrician may install an additional circuit from your switchboard, as induction cooktops typically require 20A, 32A or even 42A cables to accommodate their power demands. Most homes are unlikely to need three-phase connection, but be sure to check with an electrician before purchasing a cooktop, so that you're not surprised by unexpected installation costs.

Compatible cookware

Induction cooktops work with cookware with a ferrous base. This includes cast iron, steel, some enamelled steel, and most (but not all) stainless steel cookware. Glass, aluminium and copper generally do not work, unless fused with an iron core. If a magnet will stick to the bottom of the pot, you can use it with an induction cooktop.

Getting it installed

1. **Research**: Visit appliance stores and work out which induction cooktop you want to purchase, taking note of dimensions and power. Also research if you need to replace any of your cookware.
2. **Electrical advice and quotes**: Engage an electrician for advice on what additional circuits may need to be installed and whether your switchboard needs upgrading.

3. **Gas removal**: If you're replacing a gas cooktop, a plumber will need to come and disconnect the gas and remove the old cooktop.
4. **Install**: Your preferred electrician will make the electrical upgrades and install your new cooktop!

Government assistance

National: Under the federal government's Household Energy Upgrades Fund, you may be eligible to apply for a loan with discounted finance to afford the installation of appliances.

ACT: Under the Home Energy Support program, households in the ACT can access up to 50% off the total installed price.

TAS: Interest-free loans are available under the Energy Saver Loan Scheme.

VIC: Victorian Energy Upgrades (VEU) offers around $140 off induction cooktops for eligible properties.

This information was correct at the time of writing. Check energy.gov.au for the latest updates.

Renters

A portable induction cooktop is a great option for renters. It can sit on your kitchen bench and plug into your existing power points; no electrical upgrades or installation are required. It can be moved from house to house, is relatively cheap (less than $100), and just as energy efficient as an installed cooktop. They're typically limited to one or two elements, and have less power than an installed induction cooktop, but many happy tenants have simply popped a portable induction cooktop or two over their existing gas elements. Easy!

If you're a renter and you're stuck with an old electric resistance stove, the kind with coils on the top or an older glass-surfaced one, you're using (slightly) more electricity than you would with induction, but the most important thing is that you aren't burning fossil fuels and fouling the air in your home.

Cooking FAQs

How does the temperature control compare to cooking with gas?
Renowned chefs such as Neil Perry have made the switch to induction because of its precise temperature control. Because induction cooktops directly heat the cooking surface and not the surrounding air, the cooking temperature can be very precisely controlled.

What if I leave the burner on after cooking?
One of the benefits of induction is that the cooktop surface only heats when a pan is present; after you've finished cooking and removed the pots and pans, no heat will be generated, making it safer than gas and traditional electric cooktops.

Do I need three-phase power for an induction cooktop?
For most houses, single-phased power is sufficient. But check with an electrician before purchasing an induction cooktop, in case any electrical upgrades are required. (See Part 3 for more detail about three-phase power.)

Can I use an induction cooktop if I have a pacemaker?
It is best to check the specific appliance details and to discuss it with a healthcare professional. Induction cooktops generate electromagnetic

fields, which can affect a pacemaker if you get too close. The British Heart Foundation recommends: "Keep a distance of at least 60 centimetres (2 feet) between the stovetop and your pacemaker. Most people should be able to use an induction cooktop if they follow these precautions."

Electrifier: Kim Loo

Kim Loo is a Sydney-based general practitioner and advocate for climate solutions, serving in a dizzying number of medical, community and environmental organisations, including Asian Australians for Climate Solutions, Sydney Community Forum, the Council on the Ageing NSW advisory committee for climate change and how it impacts older Australians, and Doctors for the Environment Australia. When asked why she works for so many climate organisations at once, she replies, "Because it is a crisis, and I now have a platform."

Kim's Buddhist parents grew their own vegetables and were always concerned about saving energy. Once she had her own home and children, she started a garden and planted trees in her backyard that are now massive. She lives in Western Sydney, which is now at constant risk of extreme heat. "The trees were the cheapest way to cool down the house," she says. She also has a food forest in the backyard, with garden beds and ten chickens. She installed a wrap-around pergola covered in grape vines to keep the chickens cool.

Although her family's energy needs were already low, she installed LED lights and put blinds around the house to use less electricity. In 2015 she fully electrified the house, with a 6.4 kWh Tesla battery and 4 kW of solar panels. In 2019, she got a Nissan Leaf EV, and her two grown children also now drive EVs.

Kim has had an induction cooktop for about four years. She replaced an old electric resistance coil cooktop when it died. “When anything dies, I just get the most energy-efficient replacement,” she says. She does a lot of advocacy in her community, and one way is by showing people her garden. “I engage people who disagree with me by inviting them to meet my chickens,” she says. Chicken diplomacy! I love it.

Kim is working to educate doctors so that they, in turn, can educate their patients about the risks of gas cooktops. She talks to doctors about how people with chronic obstructive pulmonary disease and other lung issues are worse off with the emissions from gas stoves – if they can’t switch to induction, they need to be sure to keep the windows open for ventilation. “If you’ve got lung disease or heart disease, the particulates from gas stoves can actually make it worse,” she says. “The same goes for children with asthma.” This was certainly true of my own father, who experienced a remarkable lift in health when my parents removed all gas from their home.

“More doctors are understanding that they need to engage their patients and ask them the right questions,” Kim says. “If a kid has asthma, the doctor needs to ask if they’ve got a gas top or gas heaters at home.” She’d like to put up posters everywhere, spreading the message: “If you cook with gas, tell your general practitioner.”

Kim also advocates for the advantages of induction cooking to community groups. One of the goals of Asian Australians for Climate Solutions, of which she is CEO, is to convince people to get off indoor gas in order to improve the air quality in their homes. Kim has done several cooking demonstrations for community groups. “Sometimes people can’t make a transition until they see that things work,” she says. She has made spring rolls for eighty-six Chinese elders. She has cooked for the Inner West Council, which installed induction in their sustainability centre, and for the Sydney Community Forum.

Kim is infectiously optimistic. At each demonstration, she emphasises the practical benefits of induction. "It's so easy to clean," she says. It can be moved around, which is great for cramped kitchens. In summer, an induction cooktop doesn't make the kitchen hot. There's no chance of touching a hot stove or setting something on fire, which can be particularly reassuring for older people and their carers. "One day, my ageing mum turned on one gas burner, and all four went on," she says. "It's a risk." Kim's heritage is wonderfully mixed, and consequently she cooks a complicated cuisine known as Nonya, at the intersection of Chinese, Indian and Malay. I ask her which cuisine is the toughest cultural test of induction cooking, and she instantly replies, "Nonya!" She finds the cultural excuses for not using induction tiresome. It's just better. Safer. Cleaner. Faster. Cheaper. Healthier.

Australia would do well to have a Dr Kim of every cuisine, showing us we can have our incredible mix of cultural dishes, cooked cheaper, cooked faster, cooked healthier.

11
Living the Dream

Once you've electrified your home, you're going to want to make all your appliances work together, to get the best combination of comfort, cost and clean energy you can. So far, reading this book, you might have found yourself thinking, "This is all good in theory, but how does it work in the real world?"

Imagine your home as a mini power grid. You've got solar panels on the roof, an electric vehicle (or two) in the driveway, some heat pumps for water and heating, and an induction stove. The thing is, energy isn't always in sync with your needs. Solar energy might be plentiful during the day when you're at work, but you need it in the evening, when you're cooking dinner or watching TV. And if you've got a car that needs charging or a heat pump to keep the house warm, the timing and coordination of it all becomes a bit tricky.

That's where a home energy management system, or HEMS, comes in. It's like the brain of your energy ecosystem. Think of it as a smart coordinator that talks to all your devices and tells them when to use power, when to store it and when to let the grid take over. It knows when there's an excess of solar, so it'll automatically charge your car or heat your home. When there's a shortage, it'll dip into stored energy or even buy power from the grid at the most cost-effective times. It makes sure everything runs efficiently, intelligently and in a way that saves you money and reduces your carbon footprint. It can even sell your excess energy from your battery into the grid at the moment it will get you the highest price.

But a HEMS isn't just about optimisation – it's also about resilience. By integrating energy storage, you can keep your house running when the grid goes down, using your own stored energy to charge your phones and laptops and power the lights, refrigerator, stove and medical devices. It's like giving your house a survival instinct, while lowering your dependence on external sources of energy. Your home becomes a network of smart, connected devices working together. It turns every piece of your energy puzzle into an active participant in a much larger, distributed system. The future isn't just about individual homes being "off-grid"; it's about homes being "grid-smart", able to dynamically contribute to a cleaner, more resilient energy grid.

The best way to explain HEMS, and to underscore the economic story and technological story of this book, is with the experience of my mate Fred Hopley.

Electrifier: Fred Hopley

Fred Hopley is a South African expat who ran the development and operations of a media tech startup. He electrified his family's home in Sydney and is nerdy enough that he developed his own HEMS to run it cheaply and efficiently. He put solar on his roof, added a backup battery, and electrified his water and space heating, his stove and most of the family's transportation (of their three cars, two are electric, and he also rides an e-bike). I met Fred in a national park north of Sydney at my sister's fiftieth birthday party; he recognised me as the author of *The Big Switch*.

Like many parents I know, Fred was motivated to reduce his carbon footprint to lessen the impacts of climate change for his two daughters. What he has found, he says, is that "a fully electrified house is much cheaper to run, is healthier for us and the environment, and is just as comfortable". His wife and kids love driving cars on stored sunshine, and nobody misses futzing with the lawnmower.

He has saved a lot of money, too – the direct electricity savings are about $5000 a year, a number that will go up as the cost of grid electricity rises. He paid $34,000 for the solar and battery system and is on track to pay it back in seven years. There are also additional savings of $3800 a year from turning off the gas and avoiding petrol. Since electrifying their home, Fred and his family are using double the electricity they used before – but because of their solar, they only take 20% of that from the grid; the rest is from their solar panels, either directly or stored in the battery.

Fred was able to afford the upfront costs of electrification, but he believes that financing needs to be available to all households. "Unfortunately, Australian energy policy is still very much entrenched in a centralised generation and distribution framework," he says. Decentralised electricity generation through rooftop solar is much more efficient, and the Australian market is yet to catch up. "We are on an unstoppable path to a fully sustainable energy system," Fred says. "We will have abundant clean energy that is practically free for many hours of the day and often for weeks and months at a time. The economics of adding clean energy are now far more attractive than replacing fossil fuel generation."

Fred built his own HEMS to do things like charging the EVs only when there is excess solar, optimising the use of the battery, and running the pool pump when electricity is cleanest and cheapest. Thanks to the HEMS, when the family needs to buy electricity from the grid, they can take it during off-peak hours and use it to charge up their battery, saving more money. Fred hopes that this kind of HEMS will one day be easy to use and available commercially, allowing homeowners full control over their systems.

Since Fred is a fellow nerd, he was happy to share the specifics of his electrification journey. He rides a Focus Bold e-bike (13,500 kilometres so far!) and drives a Tesla Model Y. They heat their home with a ducted reverse cycle heat pump upstairs and a split system downstairs, and they cook on an induction cooktop. He has forty solar panels (15 kW) on his roof, and a 13.5 kWh battery. He times his electric pool pump to use his solar, and he has an electric lawnmower. His favourite appliance? The e-bike, which has made his commute much more fun, "rain, hail or shine".

His only regret? "Waiting too long for a technology to get better when we should've just pulled the trigger with what was available."

Fred's story is a great example of how electrification can work in the real world. He isn't a hypothetical example, and his family isn't the non-existent "average" household. Yet they are living the all-electric dream, and their real-world results can help others to know what to expect, while demonstrating what a home energy management system can do.

Let's start with Fred's solar experience. Figure 11.1 tells the story. Each dot represents the total production of Fred's solar system for a day. There are more than 1000 dots – three years' worth. You can also see his average daily electricity use over the same period, 34 kWh, as the horizontal line. Just to emphasise that there is nothing unusual going on, I have

Figure 11.1

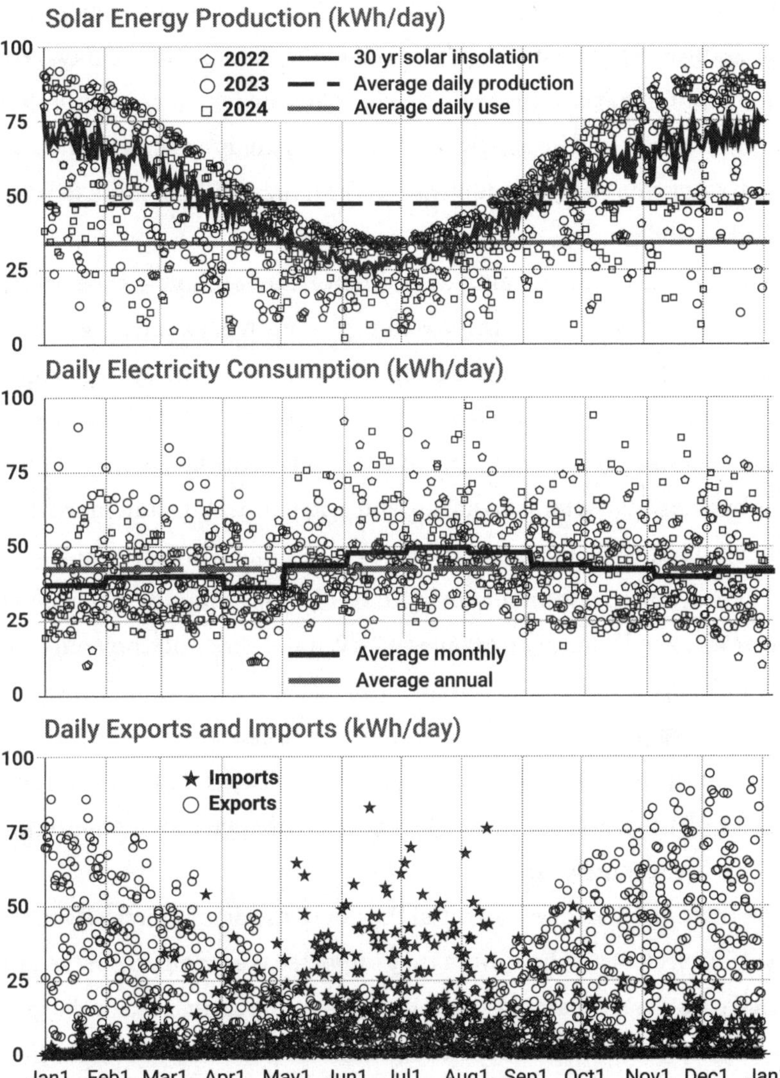

Hopley Household annual production, consumption, imports and exports. The first chart compares three years of real daily production with thirty years of data from the Bureau of Meteorology local solar sensor. The second chart shows variation in daily consumption. The final chart shows days when the household is exporting (selling solar) versus days when it is importing (buying grid electricity).

also included the average daily sunshine, measured near Fred's house by the Bureau of Meteorology. Quite amazingly, Fred's 15 kW solar system produces more electricity than his household uses on 70% of days. You can see that for most of the year, he is producing far more energy than the family needs. In winter months, when his panels are producing only about 40% of what they produce in summer, they still meet the family's demand most days.

Cold days, dark days, rainy days, blistering hot days – they are all dots on this map. Of course, there is huge variation between households, depending on where you live. If you live north of Sydney, the winter slump will be smaller; if you live south of it, it will be larger. But Fred's experience, which is backed up by the economics outlined in the first half of this book, is that solar is so cheap, you might as well install more than you need. That way, it will cover you for many days in winter, and even some cloudy days in summer, and still be cheaper than the alternative. On average, Fred's family produces 47 kWh per day, and they only use about 34 kWh.

The second part of Figure 11.1 shows daily consumption. Yes, on average, the family uses 34 kWh per day, but there is a pretty wild variation, from 10 kWH to 80 kWh. This reflects variables such as whether or not anyone is home, and whether the cars are charging.

The third graph shows how often the house exported electricity to the grid (small circles) and how often they imported electricity from the grid (small dark stars). This shows the value of staying connected to the grid, the electricity network that connects all Australians with each other. Even with a solar array that produces on average 50% more electricity than they need, Fred's household couldn't practically go off-grid without a truly enormous battery.

Figure 11.2

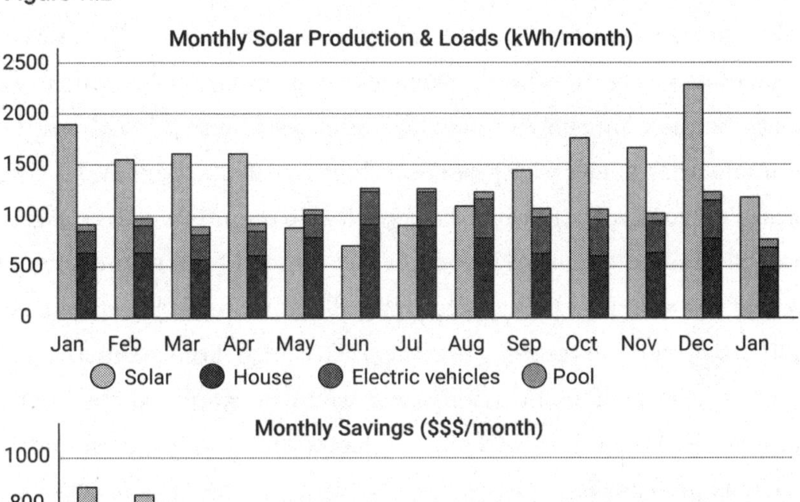

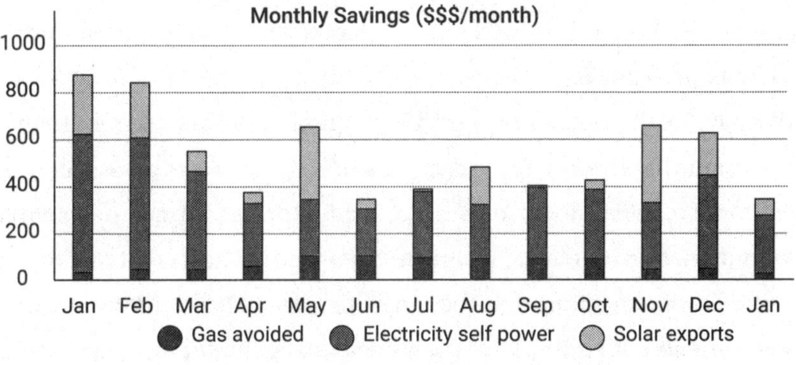

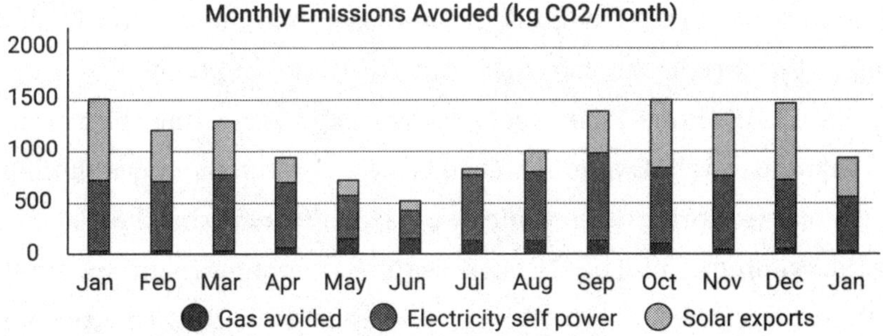

Hopley Household annual production and loads, cost savings and emissions savings. The first chart compares production versus use for the three main loads in the house: the cars, the house and the pool. The second chart shows monthly savings in dollars. The final chart shows emissions reductions.

Fred's retailer, Amber (we'll learn more about them later in the book), allows Fred to sell his electricity back to the electricity market at wholesale prices. He wouldn't advise you to do this unless you have a battery, because occasionally, for a few minutes or even a few hours, the wholesale price of electricity can be as high as $15,500 per MWh – fifty times what you typically pay for electricity. So it's only worth engaging with the wholesale market if you have a battery – both to remove the risk that you will need to buy electricity at that price, and to store electricity that you can later sell when the price is high. Fred estimates that without the battery, his electricity costs would be around $2200 higher each year. So the battery will pay for itself over six years, or sooner if electricity prices rise.

We have all this detail about the Hopley household's energy use thanks to the brains behind Fred's system – the HEMS or smart energy device. Fred works in technology, so as an early adopter he took an open source home automation tool called Home Assistant and customised it to give him an overview of everything that goes on in his system. It records everything. I am a nerd too, but I didn't have the patience to do this for my house, so I'm super thankful to Fred (and to all the developers and contributors behind Home Assistant) that he shared this data with us for this book.

The HEMS allows Fred to set parameters and priorities for heating and charging. His hot water system heats water when the sun is shining, his cars charge during the day, and he can largely avoid using the grid during the evening peak. The HEMS tells the battery how to interact with all these pieces and when to play the wholesale market. It can even look ahead to the next day's weather forecast and plan accordingly!

The details are both fascinating and encouraging. The average price Fred has paid for charging his cars is a phenomenally low $1.80 per

Figure 11.3

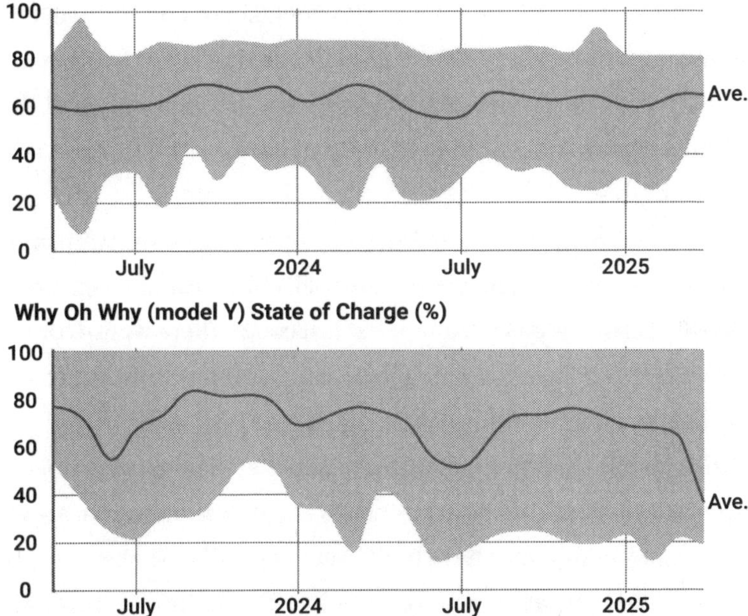

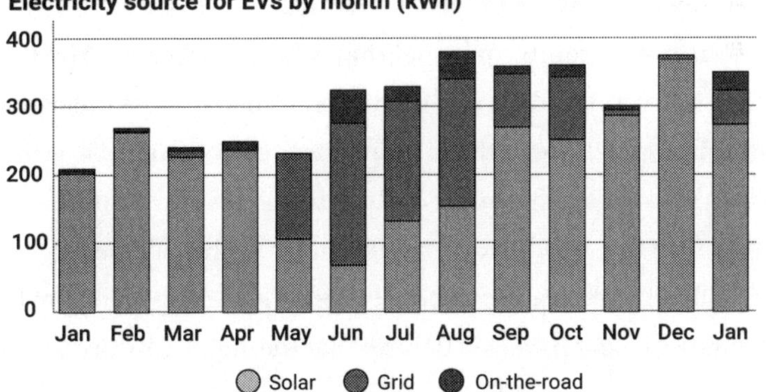

Hopley Household vehicles, state of charge and source of charge. The first two charts show the daily variation in rate of charge (grey) and the average state of charge for both electric vehicles in the household. Note that the model X has an NMC battery and the Y has an LFP, and the NMC prefers to be kept below 80% to extend its lifespan. The third chart shows the source of charging monthly: rooftop solar, grid or on the road at a fast charger.

100 kilometres, or 1.8 cents per kilometre. A similar petrol car would cost ten times as much to run.

If we get this energy transition right – which is not guaranteed, but I am hopeful – every home will have a system like Fred's. Those systems will coordinate with the larger electricity market, and not only will the system run like clockwork, but the presence of batteries on the grid will make it more reliable.

Not everyone is going to want to download open source software, modify it, glue together custom pieces of control hardware and program it, like Fred has. That said, Fred isn't alone – at last count, there were 40,000-odd people in Australia using Home Assistant. But thankfully, there are simpler-to-use products coming.

Personally, I would like to see a future where everything just connects to the system by itself, so that you don't have to think about it. We aren't there yet, but many companies have built excellent HEMS that are pretty easy to use. I like Solarhart's, as well as Schneider/Clipsal's. Tesla's is pretty good but doesn't play that nicely with others. The CEO of SPAN, a friend of mine, has an incredible product that is "set and forget". You tell it roughly what you want to do – you might tell it to prioritise charging the car on weekdays, and to keep the kitchen powered above all else – and it does it. This is what's coming, but it isn't quite here yet for everyone at every price point. In the meantime, for a few more years we'll be relying on friends, neighbours and tradies to help us set up our electrified homes.

Internet pundits would have you believe that the electric future won't work, but they are wrong: it will. As you can see from Fred's graphs, we don't really have a "baseload power" problem anymore. Instead, the big challenges are how to power the evening peak, when everyone wants dinner and a shower, and how to keep everything running in winter, when solar production drops to one-third or one-half of its summer peak in some places. A well-designed HEMS can navigate these challenges and make everything work together.

The solutions Fred has found at a household level are similar to those we will need on a national scale. We need to do, nationally, what Fred has done at home: build more solar than we need, use lots of batteries, keep households connected to the grid, and make sure the grid can keep everything running twenty-four hours a day, seven days a week, 365 days a year. At a national level, this will mean ensuring we build wind, geothermal and hydroelectricity as well as solar, so that the grid has a diverse source of energy year-round.

When all the components of the system work together, our households and our grid will be more resilient, and we will be better able to respond to climate change. As the climate heats up, we will face more extreme storms that will take out parts of the grid. In 2016, storms destroyed transmission in South Australia, and the whole state went black. Won't it be great in a few years, when millions of households have hours or days of electricity storage, enough to power their electric homes and vehicles through such events? Imagine the confidence it will give people. I like to imagine that when future disasters hit, households with backup will help those in the community who don't yet have it, increasing climate resilience across the community.

Fred's experience has made him passionate about the rules of the electricity market, which should change to reflect this new reality of widely distributed energy resources, and the important role that households will play as part of our national energy infrastructure.

Which brings us to our next section. As well as updating our homes to run on clean electricity, we need to update how we buy, sell and regulate electricity, so that the market rewards people for electrifying and reflects the realities of our new grid. In Part 3, I'll explain how the market works today, how it got that way, and how we can make it fairer for everyone.

Part 3

Buying and Selling Electricity

12
A Brief History of the Grid

Before we look more closely at how our current energy market works and how we can make it fairer and cleaner for everybody, it's useful to understand how it came about in the first place.

In the earliest days of electricity, a local generator – probably running on coal – made electricity somewhere in your city. It would connect to all the homes and businesses in a local grid that extended (if you were lucky) to the edges of town. Towns grew and electricity became more popular, so eventually the grids started to overlap. (At one point in New York, the birthplace of commercial electricity networks, there were nine separate networks of poles and wires.) Rather than run separate sets of infrastructure, it made sense for these networks to join together.

The town of Tamworth in New South Wales had the honour of being the first community in Australia to provide electricity as a public service: in 1888 it energised a network of streetlights with a small generator. Other towns and suburbs followed across the country. An interconnected transmission grid was formed, usually covering an entire state. Local electricity supply was left to local distribution companies, many of which came to be owned by municipalities.

In the twentieth century, state governments moved into electricity to support growing populations, manage colliding municipal networks, and keep up with the massive first wave of electrification across industries, from shearing to smelting. After WWII, the federal and state governments stepped up electricity generation, because a tapestry of local power stations was never going to power the energy demands of a nation that was growing rapidly and industrialising. In 1949, the Snowy Hydro Scheme was established, and electrification was well and truly a national priority.

Until the 1980s, cities such as Wollongong, Newcastle and parts of Sydney and Melbourne operated their own electricity networks. It didn't make sense to have two sets of infrastructure, so there was just one – a "natural monopoly", as economists would call it – and it was owned by the city. Public servants kept it running and made sure prices didn't get out of hand. It helped that technology was improving: with scale, electricity got cheaper during this period. In exchange for their right to operate the monopoly, cities were required to guarantee they wouldn't disconnect at-risk customers.

In the late 1980s, it was deemed a good idea to roll up all the remaining local government electricity providers into the existing state government-run entities, corporatise them and sell them to private investors. Even the ones remaining in public hands, previously known as "commissions", became "corporations", ostensibly delivering public goods more efficiently by working more like private enterprises, maximising value for the shareholder – that is, the public. Fast forward to today, and some of these publicly owned companies are making higher returns for their shareholders by price-gouging consumers. It is a system that only a big consulting firm or investment bank could have come up with.

In the 1990s, when electricity prices started to rise, it was decided that we needed a National Electricity Market. People trained in "market design" tried to achieve two things at once: protect households from price-gouging monopolies and create "competition", which, according to the neoliberal ideology of the time, would magically lower prices. This is how we went from a "vertically integrated" monopoly that did everything – from generating the electricity to managing the local wires and even sending you the bill – to the current Australian energy market.

In this new, "competitive" market, we had generators (who made electricity), transmission companies (who moved it long distances across the country), distribution companies (who got it to households and businesses) and finally retailers. Retailers did two things. They traded electricity on the wholesale market, and then they packaged it up through a retail "plan" and sold it to consumers.

Pretty much all these developments happened before the mass adoption of rooftop solar and batteries, the acceptance of climate change as a genuine threat, and electrification as a solution. As such, the system, including the market, is not designed for today's reality. The old system was designed for electricity to flow one way, from a generator, such as a coal-powered plant, across the transmission system, into the local distribution system and across the meter, where it was measured and billed to you. Most of the electricity generated could be seen as being part of a single, centralised "pool". Electricity flowed into the pool from generators and then flowed out to be used by consumers.

Today, with the shift to rooftop solar, households have become one of the largest generators of electricity, yet they get piddling compensation for selling electricity back to a network they once owned. The design and regulation of the system continue to favour investors over taxpayers.

Australians already have over 4 million solar systems and will soon have more, as well as batteries and EVs and all the associated kit. The question is: how should we design the market for this new world? Households have become the biggest and most reliable investor in the National Electricity Market, but they have no rights as investors. It doesn't sound like a level playing field to me.

To get a sense of some of the absurdities that have resulted from using an outdated system, consider the city of Newcastle. I was recently there and heard the deputy mayor regret the moment in the 1980s when the electricity grid was taken away from the city's control and put under state control. The city of Newcastle has a bunch of facilities: bus depots, offices, parks and so on. It is now discouraged from putting solar on the roofs of these facilities because, instead of being able to use that electricity at other sites on the same distribution grid, it is forced by "competitive market rules" to sell it to the National Electricity Market. The city is paid a modest feed-in tariff for this electricity – and then the city has to purchase electricity back from the grid at retail prices. It could be using its own solar at a cost of 5 cents per kWh, but instead it ends up paying 30 cents per kWh or more. This costs you, the taxpayer, money, and gives that money to the private entities, such as Macquarie Bank and foreign investment firms, that engineered this system.

Competitive market my arse

Because there are natural monopolies in the system, the "market design" was meant to prevent price-gouging. Rather than gouge on price, however, the grid operators now gouge on the "regulated asset base", or RAB. The grid operators are guaranteed a profitable rate of return on any given investment. The Australian Energy Regulator (AER) is meant to act as a watchdog over these investments, but it isn't given enough information to really keep a lid on "gold plating" – the practice of over-investing to take advantage of the guaranteed rate of return. Over-investment by grid operators often leads to higher prices for consumers.

The Australian Energy Market Commission (AEMC) makes and revises the rules for the electricity market, while the Australian Energy Market Operator (AEMO) runs it. Back in the early 2000s, it was seen as a good idea to allow generators and retailers to combine into super-businesses called "gentailers". So now, in this purportedly "competitive" electricity market, the generator side of a business can jack up wholesale prices, and then the retail side of the same business passes the inflated prices directly to customers.

Large retailers also use a "lazy tax" to maximise their take from consumers. If a customer dutifully pays their bills regularly, their provider rewards them for their loyalty by hiking up prices. When consumers complain, the industry and regulators tell them they should have shopped around, essentially putting the blame for corporate greed onto the consumers who are its victims.

All of this is to say that the market wasn't designed for you, or for the twenty-first century, or for the reality that our rooftops, cars, batteries and appliances are now the most important and largest contributors to the grid. The twentieth-century model of a regulated monopoly is failing in the twenty-first century.

Why do I say all this in a book about how to electrify your home? Because the rules of this market are going to determine who the economic winners are. Currently, Australian households do not have a voting seat on the market bodies I described above. The people sitting on those boards work for energy companies, banks and infrastructure companies, and their interests are not aligned with yours. We are going to have to fight for our place at the table.

Today's market has made it illegal to run a cord over your fence and sell your electricity to your neighbour. It would like to pretend, when it buys your rooftop solar for a piddling sum, that this electricity then flows all the way back to some central "pool" – and that the electricity you buy from the grid, at many times the price you received for your solar, has come all the way from that same distant pool.

In reality, your excess electricity will most likely be absorbed by a nearby neighbour. Physics and your own intuition will tell you this. It is a consequence of Kirchhoff's laws of electric circuits that electricity will flow to the nearest loads – that is, to your neighbours. Indeed, the more electricity we share over the local wires, the cheaper those wires are to run. A fair and efficient system would maximise the local generation and sharing of electricity, creating the cheapest possible energy system for everyone.

Some of the benefits of this new dispersed generation network should go to households. That much is obvious. But there are other questions about how bills should be reconciled – should, for example, local exchanges of electricity be balanced before households buy centrally generated electricity from the pool, or should everyone play the wholesale market? These are questions we still need to nut out – and households must have a say in these decisions.

This may all sound abstract, but it will determine whether or not we have cheap energy and fair rules that give everyone a fair go. Reform only happens every fifty-odd years in electricity markets. We have a chance to do it right now.

For twenty years, Australia's solar inventors and leaders have been calling for a simple market design for the solar age: price parity. Our solar scientists, who did much of the heavy lifting to create the silicon PV industry, also foresaw that the old coal-centred market would throttle its new clean-energy competitor. They never asked to lock in vast tariff handouts for households – they just proposed that solar households should get paid what the generators get paid.

Here is my updated manifesto for a revolution in the energy market:

1. No one's capital expenditures should be advantaged over anyone else's. Individual households and investor-backed networks are equals.
2. The cost of electricity should reflect the physics of electricity, rewarding locality and minimising costs for consumers.
3. A watchdog with market teeth that represents the end consumer should be built into any monopoly market. Consumer representatives rather than corporations should comprise the majority of its board of directors.
4. Consumers should have a voting representative on all market bodies that regulate, design or oversee the electricity grid and electricity markets. This representative should be independent of any political parties.

Australia has the chance not only to lead technologically, and to atone for its previous climate sins with climate success, but the chance to show the world how to build a fairer, more democratic, and lower-cost energy system for everyone.

Electrify everything for everyone, everywhere.

13 Understanding Your Electricity Bill

Buying clean electricity from the grid and selling the electricity you make with your solar panels back to the grid have both been made unnecessarily complicated. Even experts find it hard to read an energy bill in this country! For some help explaining how electricity is bought and sold, and how households can buy and sell electricity most effectively, I turned to my mate Dan Adams.

Dan was originally an aerospace engineer and spent time designing wind farms. His experience reflects my own: when we started our careers, we worked on renewables technology because renewables needed to be cheaper. That isn't the problem anymore. Renewables are now very cheap. The problem now is how to integrate them into the grid and how to match supply (where electricity comes from) and demand (where it is used). Dan's view is that automating batteries, EVs and other devices in the home is the best way to do that. I strongly agree.

After working for some years at Tesla, Dan is now CEO of Amber, a new energy retailer in Australia that is helping Australian households to get the maximum value from their electrification investments. There are better options out there than buying and selling to the grid at whatever price the utility companies happen to set. Instead, Amber enables owners of batteries, solar, vehicle-to-grid EVs and other assets to trade their energy in the wholesale energy market.

When the price is high, Dan helps customers to sell the energy on their roofs or in their batteries at that high price. Since much of the energy infrastructure of the future is distributed across the country and in people's homes – on roofs, in driveways – both Dan and I believe that the compensation and incentives for hosting this infrastructure should be widely distributed, too. This new infrastructure should be as well supported and as fairly compensated as the more centralised infrastructure of the past, such as coal and gas plants and their big transmission lines.

I asked Dan to explain my electricity bill to me. Making sense of what you're being charged can be daunting. I recently moved house, and in all the madness of that, I hadn't even looked at one of my new bills yet. Dan explained that it is normal to find your bills confusing; the big retailers make them complicated because they can price-gouge confused consumers. The business model of a traditional electricity retailer is to get you in at a cheap price – promoted through an energy comparison website or sneaky advertising – and then raise the price without you understanding what is happening. In order for this rip-off to work, they make the bills complicated, so that you stop paying attention. The price slowly jacks up while you're not looking. Not unlike that Netflix subscription, you forget that they make most of their money from the customers who don't engage at all. I literally write books about electricity and electrification, and I've been one of these ignorant customers. I get bored of it too, and want to give up – but for you, dear reader, I push on.

Electricity regulators, federal governments and big retailers created a system without price controls. Instead, consumers were supposed to shop around to avoid being ripped off. Some states did push back and provide protections, but at the national level there was no honest grappling with the fact that most people either do not want to shop around for electricity, or don't know how to.

Our first electricity bill after we moved house came to a whopping $1167.14. It took me some searching to find out what period this covered: eighty-six days, I finally discovered in the fine print. That means I'd been spending about $400 per month, or $13 per day, on electricity. "That's high," I thought, and then immediately felt stupid that I hadn't looked at an energy bill sooner. I was spending $4000 per year on electricity, while still buying a little gas and a little diesel. My gas bill was $146.23 for the last quarter, or around $600 per year. Most of our driving was done in our electric BYD Dolphin, but I'd still spent about $2200 at the pump for various road trips in our diesel-powered Toyota HiAce bus. All up, I had paid close to $7000 for energy in 2024. This is about the Australian average, which makes sense, since we are a family of four in an old, average-sized house.

Electricity-wise, I had bought 3413 kWh at a rate of $0.2819 per kWh. The small existing solar on my roof sold 108 kWh back to the grid at $0.07 per kWh on an increasingly diminishing feed-in tariff, for a total credit of $7.56. I paid $0.78 per day for the connection to the network, or $285 for the year. Gas-wise, my bill was $146.23, nearly half of which was for the connection fee, at $0.78 per day! We had electrified our cooking and put in a reverse cycle heater, so our only remaining gas use was the ageing in-line water heater. It therefore made sense that our gas use was low. (Neither I nor the kids take many showers, as we consider a dip in the ocean good enough. My wife is more civilised.)

Given my high electricity bill, I asked Dan to talk me through my retail plan options. He explained that most people have three options, depending on their circumstances.

If you use a lot of power during the evening peak – roughly 5 p.m. to 8 p.m. – you will be better off with a traditional electricity retail bill through a traditional electricity retailer. This is often called an "anytime" tariff: you'll be charged the same price at all hours of the day, meaning you are not charged extra during the evening peak.

If you use a lot of power in the middle of the day, or if you can shift a bunch of your usage to that period, you probably want to be on a "dynamic" tariff, which means the price varies throughout the day, with lower prices in the off-peak hours. The simplest dynamic option is a "time-of-use" tariff, which you can get through traditional retailers.

The third option is something like Amber, whereby you're basically buying at the true wholesale energy price as it varies throughout the day. In the middle of the day, you'll have access to lots of cheap renewables. This is the cheapest possible way to buy power during off-peak periods. During the evening, when power is coming from coal and gas, it will be more expensive. If you have a battery, or if you can shift more of your usage to the middle of the day, this should be the cheapest way to buy power, and you will be supporting the energy transition by using more renewables. If you have a battery or an EV plugged into the grid, you'll be able to discharge in the evening, meaning you'll get paid to displace fossil fuels.

When Dan worked at Tesla, he developed their virtual power plant (VPP). A VPP is a program whereby people sign up to have someone else manage their battery for a fee. Dan learned that customers don't really want the VPP model. We are a little way into this experiment, and 86% of households with a home battery in Australia are opting out of VPP programs. Dan explains that the reason is simple. People want two things: to maximise the value of their battery and solar, and to stay in control of these assets. Unsurprisingly, people hate giving up control to their utility company.

Instead, customers want a model that will help them to unlock the full value of their batteries and EVs while giving them control.

This is what Amber has tried to provide, Dan explains. "Our whole model, and the reason we've done things the way we have, is so that an individual customer can get paid the same price as a big coal or gas generator when they're exporting their battery into the grid, and they can buy power at the same price as big retail when there's lots of cheap renewables flooding the grid," says Dan. "We basically empower the individual household to be able to compete with the big end of town."

Dan has about 38,000 customers so far. Most of them come to Amber, he explains, after buying a home battery system and talking to their tradie about it. Increasingly, the tradie tells them they have three options. One: they can just use their battery for self-consumption. Two: they could sign up to a VPP, which might pay them $300–400 per year for the electricity they sell back to the grid, but they'll lose control of their asset. Or three: they can sign up to a company like Amber, get direct access to the market, get paid the same price as a big coal or gas generator, unlock about $1000 per year of additional value and stay in full control. Many customers, like Fred earlier in the book, see it as a no-brainer. Once again, tradies are the climate heroes and the frontline salesforce for this all-electric world.

Dan's vision for the future is to see at least 5 million home batteries and EVs in Australia by 2030. Those batteries and EVs will have enough storage capacity to meet most of the storage requirements for the energy transition. We will also need some long-duration seasonal storage, such as Snowy 2.0's pumped hydro storage facility, but the heavy lifting will be done by batteries. Five million home batteries and EVs equates to around 30 gigawatts of capacity. They could pump thirty large coal-fired power stations' worth of power into the grid for a few hours at a time.

The Australian Energy Market Operator (AEMO) predicts that we will need about 18 gigawatts of battery storage to enable us to get to 82% renewables. You can't unlock 100% of people's home and car battery capacity – but if you can unlock even half of it, you're a very long way towards meeting the storage requirements of the energy transition. Dan and I agree that this is probably the cheapest and fastest way to get there, because it involves batteries and EVs that people are already buying, rather than paying some monopoly generator to build large-scale batteries over a hill somewhere.

This is the future. It'd be crazy if we weren't using those batteries to get more renewables into the grid. But the question remains as to whether we have the political will to reform the electricity market. My view is that we should fix the incentives in the electricity market so that Australians are motivated to electrify everything and gain the full value of their assets – in the same way the utilities are currently incentivised to maximise the value of the old infrastructure. Let's level the playing field.

Some good news here is that late in 2024, Clare Savage, Chair of the Australian Energy Regulator (AER), in part because of community pressure created by Rewiring Australia and others, announced that the AER would be looking at the current rules of the grid and allowing new rules to be tested in regulatory sandboxes.

Until now, taking control of your electricity use has been hard. You had to install a smart switch on your old hot water system or pool pump. Or you needed to call in an expert. Or you found that your wi-fi didn't reach the system. Now, the new batteries and EVs connect to the internet, and there's an interface you can use to control them. We no longer need to install bespoke hardware. The future is finally here. As the *Fully Charged* podcast put it, "the energy transition is no longer about technology; it's about psychology". By buying solar and an EV that feeds its battery back to the grid, people will be able to have a $0 petrol bill and a negative energy bill. That is Dan's goal. (Talking to Dan reminds me that I'm on track for a $4000 electricity bill if I don't install solar and a battery and change my retailer soon.)

It sounds too good to be true, so I ask Dan the hard question: how on earth does Amber compete with the bigger utility companies?

"We get exactly the same price," he explains. "So we buy power. Every retailer in the market buys power at the wholesale energy price, which varies throughout the day. Prices are set every five minutes based on supply and demand. We buy power based on those prices. We pass it directly to customers – no markup. So, we just charge a fee." Amber's customers pay a monthly $22 subscription fee. "We don't make any money on selling energy, just from the subscription," Dan says. "We want to be aligned with our customers. We want our customers to get the lowest possible bill and use the most renewables possible."

When I remark that I'll have to get $264 a year worth of value out of Amber to make the monthly subscription cost worthwhile, he says, "No, because you're already paying more than $264 in margin to your utility company on your usage. The way it works is, they buy power at a variable price. Then they add their hedging costs." This hedge is the buffer that the retailer adds to make sure they are profitable. They choose it conservatively – safely and in their favour – and that guarantees them more profit than you might like them to have. Dan continues, "They turn that variable price into a fixed price. Then they add their margin, and they sell power to you at a fixed price. So they make $300 or $400 a year on you, and you'll be paying for their hedging costs as well. You don't need that hedge if you've got a battery or an EV – you have a physical hedge in your house."

I'm sold.

This future is going to happen fast, already faster than the market has been able to adjust. Dan points to Tibber, a company operating in Norway and Sweden, which does the same thing as Amber – it gives customers access to dynamic energy prices throughout the day and helps people to smartly charge their EVs. They've grown from zero to half a million customers in about five years. During the same period, EV sales went from 10% of new car sales to 90%. People found that when they bought an EV, their energy bills doubled. So they thought, "Shit, what can I do to bring down my bill?" And they signed up to Tibber. Dan and I agree that in some form or other, this will happen quickly, all over the world – speeding up decarbonisation, making the market fair for everyone, and lowering energy costs.

I put Dan on the spot about the electrification of his life. Like me and all the other "experts" in this book, he's not finished yet. Dan has rented an apartment for the past twenty years and hasn't been able to do any of the stuff he helps his customers with – no solar, no battery. He has a beat-up Subaru Outback that's twenty years old. But he has just moved out of Melbourne and bought a place. He's excited because he is on the same journey as me, with a new home and the opportunity to electrify everything and use his own products. Earlier that week, he'd had a solar and battery installer out to give him quotes. He had just put in an order for his first EV, a Tesla. The rest of his home is already all-electric, which was one of his search criteria. The kitchen has an electric stove; not induction, but "good enough for now". The house has reverse cycle heating and cooling and electric hot water. He's looking forward to signing up for Amber soon, once the battery he has ordered arrives. Dan is electrifying everything and helping the rest of us to do the same.

14
Building a Fairer Electricity Market

Australia has reached a tipping point: electrification will now save nearly every household money, even once the cost of the transition is taken into account. The economics will drive many households to electrify themselves, but for some – especially low-income households, renters, apartment dwellers and residents of social housing – reforms and programs are needed to ensure they aren't left behind, paying escalating fossil fuel bills that they can't afford but can't avoid. In this chapter, I outline the key policy changes that will make electrification work for everyone.

The finance to electrify everything can be delivered in multiple ways; the key is that it needs to be an easy "yes" for every Australian, no matter their financial means. If we don't ensure this, people will be left behind. As electrification gathers pace, the gas network, gas equipment and even petrol and diesel supply chains will become increasingly expensive and fragile. The people still reliant on them will bear the brunt of these escalating costs.

To electrify everything for everyone, we need:

- Finance that everyone can access to manage the upfront costs of electrification
- Minimum standards and disclosure requirements for rental properties
- Funding for social housing upgrades
- Programs to incentivise and support strata electrification.

Something I have always loved about this country is the idea of giving everyone a "fair go". What follows are policy prescriptions to keep Australia's energy transition in line with this egalitarian ethos.

Our finance system favours fossil fuels

We have seen that a $10,000 solar installation at the start of a mortgage will save that household about $30,000 by the end of their mortgage. Yet getting $10,000 added to your mortgage limit can be very difficult, even if, thanks to the money you'll save on electricity, it would clearly make the mortgage easier and less risky to repay. Because investments such as solar panels will save households significantly more in the long term than they cost, investing in solar actually lowers a household's financial risk profile, as the household will be less exposed to volatile fossil fuel prices.

The same is true of vehicles. If you are approved for $30,000 of vehicle finance, and an electric vehicle will save you $10,000 over the finance period, you should be approved for a $40,000 electric vehicle or $30,000 petrol vehicle. Yet this isn't the case in banking today, because banks more or less ignore energy economics. This means our financing system fundamentally favours fossil fuels, where upfront costs are low and long-term costs are high and more volatile. Electrification – whether through solar, a battery, an electric vehicle or a heat pump – involves paying more upfront to gain an overall better financial position over time. Even large-scale renewable projects are in this position.

Electrification for renters

Renters can get an electric car. They can ride an electric bike. They can even get an induction burner and use that instead of the gas stove. But these are band-aids, not structural solutions to the challenges of decarbonising your life if you are a renter.

The rental market presents a major challenge for electrification. This is because of the "split incentive" problem: the landlord chooses and pays for appliances, but it's the tenant who would benefit from lower electricity bills, meaning landlords may see little financial reason to upgrade their properties. Finance schemes can help to remove upfront cost barriers, especially for low-income landlords, but other reforms are required to electrify rentals at scale and speed.

We need minimum standards for rental properties, so that landlords are required to electrify their properties over time. This should include a requirement that landlords install electrified appliances whenever an appliance is replaced (this is already proposed in Victoria for heating and hot water systems). We also need a clear process by which renters can request and access solar panels and battery upgrades.

Another key opportunity to drive longer-term improvements is requiring clear disclosure to potential tenants about a property's electrification status, current energy bills and overall quality. Disclosing the real running costs of rental properties when they are advertised means landlords who make bill-reducing investments will be rewarded, as their properties will be more competitive.

A range of other policies could also help to drive upgrades in rental properties:

- adjusting federal tax incentives and deductions to encourage electrification
- using levies and discounts on state land taxes to encourage electrification
- requiring landlords to cover the cost of maintaining a gas connection to their rental properties, as they are the people with the power to remove it.

Electrifying social housing

Many of Australia's most disadvantaged households are tenants in public and community housing. Like private renters, they can't electrify themselves; they rely on their public or community housing provider. The solution is straightforward: the responsibility falls to governments at all levels to directly fund the upgrade of this housing.

The economic and other benefits for these households – in lower living costs and better health outcomes – are just as strong, or stronger, as they are for Australians more broadly.* They should be at the forefront of homes being upgraded. The job of upgrading our social housing provides an opportunity for the government to directly build capacity in the

* See ACOSS and Deloitte Access Economics, *Powering Progress: Energy Upgrades to Low-Income Housing*, July 2024, for more detail on the task to improve low-income housing.

supply chain and the workforce for housing retrofits everywhere.

Electrifying apartments

Strata schemes pose specific challenges to electrification. There's less room for solar, and sometimes there are more complex space or access constraints when it comes to changing appliances, plumbing or wiring. Often, significant upgrades need to be coordinated with the owners' committee – whether adding shared solar, electrifying shared services such as hot water, adding EV charging to parking areas, or paying for related upgrades such as power supply improvements.

Supporting strata homes to electrify will require a careful set of reforms and incentives. Making flexible finance available to strata property owners (including landlords, who own more than half of strata properties) is one crucial factor that will help strata groups to co-ordinate upgrades while minimising hardship or conflict.

Interest-free and lower-interest loans

Conventional loans offered at 0% interest provide a clear encouragement to households to invest and are very attractive to households that can afford the regular repayments, which will usually be lower than the resulting bill savings. The ACT has already made over 17,000 interest-free loans to upgrade homes under the Sustainable Household Scheme, and a similar scheme is open in Tasmania. The investment required from the government to provide the interest subsidy can be justified by the savings and benefits on offer, which are thousands of dollars per year per household, or a whopping $1.7 trillion economy-wide by 2050.

There are also discount-rate loans now available Australia-wide through the recently created Home Energy Upgrade Fund (HEUF) – something

Rewiring Australia lobbied heavily in favour of. This fund gives banks and other lenders access to cheaper capital, which they can offer to their customers for electrification upgrades in the form of reduced-rate personal loans, and even lower rates for people with an existing home loan. It's a good start! We're watching developments closely and continuing to lobby for a wider range of options to make finance accessible to everyone.

These loans – both 0% and discounted interest – may still need additional policy support to include low-income households, who might otherwise be hesitant to commit to the regular repayments or to engage with finance products.

On-bill finance

"On-bill" finance is a well-established idea being championed in the USA, whereby your utility company covers the upfront cost of an upgrade, and you repay it over time via your bill. Advocates consider this a "natural" way to pay off electrification, because your power bill is where electrification saves you money. Your energy bills still become lower overall, with a portion of each payment going towards financing the electric things and the rest going towards paying for the energy you've used. Mobile phone companies already do something very similar. You can get a new handset that you then pay off over time via your phone bill. In the case of electrification, it's even better, because the upgrades you're paying for will save you money.

This approach may also help to remove the barriers for renters, especially to installing solar, as the finance would be attached to the meter and paid off gradually by whoever is living in the home. If one tenant moves out and another moves in, nobody has lost out by paying for the entire upgrade upfront. In rented properties, this type of loan is best suited to solar and battery upgrades, since it shouldn't fall to the renter to pay off replacement appliances such as stoves or water heaters – these should remain the landlord's responsibility. These loans should also be repaid over a long term, such as fifteen years, so that the repayments are lower than the bill savings and the bill-payer (the tenant) is still better off overall.

There are reasons why this approach will be tricky to deploy in Australia, but to accelerate climate action we should consider all ideas. In (most of) the USA it works because the utility is a regulated monopoly, and customers don't have the choice to switch retailers. In Australia, consumers can choose from a wide range of retailers. Given this, on-bill finance would need to be portable between retailers to preserve choice and competition. There would be many questions to nut out. Should the loan stay with the old retailer or automatically move to the new one? Who pays the loan if a property is empty? There are rules about the collection of unpaid energy bills to protect vulnerable households from having their electricity cut off, and mixing the consumer protections for credit and finance with consumer protections for energy gives a lot of people (especially lawyers) heartburn. But we shouldn't write off potentially good ideas just because they are hard.

"On-bill" loans could also potentially be held by the network providers (known as DNSPs these are the companies that own the local infrastructure – poles and wires – on your street), and passed on through the network fees they charge to retailers. This could be linked to a "no higher bills" guarantee, requiring that the repayment amount built into those fees still leads to lower overall bills thanks to solar and efficiency savings. This is the basis of an impressive initiative, Inclusive Utility Investment, proposed in the USA. But this would be a big change to how Australian network providers' fees are currently calculated and passed on by retailers.

For these reasons and more, on-bill finance isn't happening in a meaningful way in Australia at the moment. Energy retailers have instead started encouraging people towards the discounted loans already on the market.

On-rates finance

Another approach that has been used in the USA and elsewhere is for property owners to repay electrification loans through their council rates.

This approach has a few benefits: it uses an existing system, it's targeted to property owners, including landlords, and it could carry across ownership – the loan could stay with the property if the property changes hands, removing the reluctance some people might have to upgrade their property before selling. The downside of this is that the cost of shoddy upgrades could be passed on to some future owner – so there's still probably a strong case for making someone pay off the loan when the property sells, with the upgrades included in the sale price and no surprises.

The main drawback to this approach is that Australia has about 570 local governments, and while some of them are right at the frontier of climate action and capable of adapting their billing systems quickly, some of them ... aren't. An expanded national approach to financing could use local councils as one option to deliver and collect loans, but federal and state governments are probably better suited to the task.

The Electrify Everything Loan Scheme

When Rewiring Australia was challenged to come up with a nation-building idea that would work at scale, we developed the concept of the Electrify Everything Loan Scheme finance that could truly work to decarbonise all our housing stock within one generation.

As we have seen, electrification at the household level has to be done by property owners. This fact is a stark reminder of the inequality in Australia's housing system, but fortunately, we can also take advantage of it.

Government, or a government-backed lender, could offer a loan to pay for electrification upgrades that is:

- secured on the property title
- indexed to inflation, and
- repaid on the sale of the property.

This would offer a simple mechanism for everyone, regardless of income, to make the switch, while still putting the bulk of the eventual cost on property owners and minimising the cost to government. If a loan only has to be repaid on the day your property is sold, it won't affect your cashflow until then – and the day you get the proceeds from selling a property is usually one of the easier days in your life to pay off a loan of a few thousand dollars!

When we first worked on this idea, we were inspired by the HECS system of student loans. HECS repayments are contingent on income, and only kick in once you are earning over a certain threshold. Your HECS debt creeps up in line with inflation in the meantime. Electrification loans could be similarly income-contingent, which might encourage people to pay back some or all of their loan early – which, based on our modelling, reduces the cost of the scheme a lot, because it recovers the loan from younger, early-career home buyers who might not sell again for fifty years. But the "pay back on sale" option is still a crucial ingredient for an electrification loan, because this loan would also need to work for large numbers of older people who own property, are retired or close to retirement, and are much less likely to have a regular income.

It would also enable landlords, even those on low incomes, to invest in electrification. Crucially, this could make it politically easier for governments to implement minimum standards for rental properties, which would deliver the cost-of-living benefits of electrification to renters too.

A bank might offer a loan like this, but this "pay back on sale" scheme relies on something governments can deploy that most financial institutions can't: a lot of patience. The average property in Australia is held for eighteen years at a time, which is a longer (and more variable) period than most banks want to wait around for.

Governments, however, can wait, and the fiscal cost of that patience is tiny. Governments can borrow money via indexed bonds that track the inflation rate, meaning the cost to the government budget would be just a fraction of a percent per year in interest subsidies. The returns to households in bill savings and to the productivity of the economy would be orders of magnitude higher – never mind the emissions reductions!

Past governments have created universal healthcare through Medicare and universal tertiary education through HECS student loans. The Electrify Everything Loan Scheme (EELS) has been a collaboration of Rewiring Australia, the Australian National University and Professor Bruce Chapman, architect of the original HECS scheme. Like education, electrification is an upfront investment in the future, with both private and public benefits. If it can be financed in a way that allows people to repay when they're best able to, while keeping the costs on property owners, it becomes fairer and more accessible for everyone.

There's another important benefit to a flexible, property-based approach that doesn't require "proof of income" paperwork: it can be pretty much issued on the spot. This is important, because about 80% of hot water replacements are done by people replacing a broken hot water system in a hurry. The plumber needs to be able to offer finance on the spot – "Do you want to pay a thousand bucks for a new gas system, or upgrade to electric, save money on your bills, and put it on your EELS account?"

One way or another, we need more finance options

By the time they reach the end of this book, I hope everyone realises just how critical finance is to a fast and fair energy transition. I'll be blunt: one of the reasons I am writing this book is to encourage an army of advocates to push governments of all political stripes to prioritise cheaper and more accessible finance. Without it, we won't hit our climate targets, and I believe we will set up a culture war. A swathe of poorer households will be left trapped in their expensive subscriptions to fossil fuels. We can avoid this by tuning our finance systems to meet the needs of everyday people.

An idea: symmetric tariffs

Right now, tariffs are "asymmetric": consumers aren't getting a level playing field to use their own investments to help the grid. For example, the grid company usually charges higher fees at peak times such as early evening but doesn't offer a "symmetric" reward to your next-door neighbour for discharging energy at 6 p.m. and reducing the peak for your street. The current tariff system pretends that when your neighbour sells electricity to the grid, it flows all the way back to the headquarters of your retailer, and then back to you when you buy electricity from the grid. In reality, electricity moves only as far as it needs to, so the electrons from your neighbour's solar rarely leave the neighbourhood. Reflecting on this, and on the fact batteries now exist, many experts now believe that we should have a symmetric tariff. The price paid to people selling electricity should match the price paid by people buying it. This would incentivise people to install more batteries and would almost certainly lead to lower electricity prices for those who can afford to participate.

A related idea: local user charges

Symmetric tariffs should probably also compensate the companies who maintain the local poles and wires (the DNSP) for moving electrons between houses and keeping the network running. This compensation is sometimes called a local use of system tariff, or LUOS. It should probably only be a cent or two per kWh, so that the household selling the electricity would make money, the network would make a little money, and your neighbour, who might be a renter or live in strata, would get cheaper electricity. This would also, in turn, create an incentive to install neighbourhood or shared strata batteries that can buy and sell to nearby houses at lower rates than big generators or storage far away. If we need fewer poles and wires, the system will be lower cost for everyone – a rare case of everyone being a winner.

With market reform and better tariff design, we will get cheaper electricity in Australia. By making better use of local poles and wires and using more of our sunshine, we can keep more money in your pocket and in your community. Opportunities to reform a utility market do not come along often. The last major reform of our electricity market occurred at the turn of the century, when utilities were privatised. That reform failed to lower prices or introduce competition; instead, it led to large cost increases. It might be too late to un-privatise the energy market, but it isn't too late to design a sensible system that gives Australians cheaper and 100% clean electricity.

Part 4

Frequently Asked Questions

We've focused in this book on how to go about electrifying the Five Big Things, make them work together, buy and sell electricity and build a more equitable and effective energy market. Of course, many other useful details will inform your electrification journey, and that's what we'll explore in this section.

FAQS

How much do I need to know about electricity?

You'd think, as Australians, we'd all understand the differences between AC and DC, given so many of us grew up on AC/DC! But we don't necessarily; the language of electricity can be confusing. What does it all mean? What is the bare minimum you need to know? And will your current electricity supply be enough?

Power vs Energy

Energy is the capacity to do a task, or "work" as physicists call it, and is commonly measured in kilowatt hours (kWh). Power is the rate at which energy is spent and is measured in watts.

A machine that needs 1 kW of power to operate uses 1 kWh of energy per hour of operation.

Electricity, in the form that is delivered to your house from the grid, comes as "alternating current" (AC). The electricity that comes out of a battery, by contrast, is "direct current" (DC). You don't need to remember this explanation, but in case you want to know: in AC power, electrons switch directions, moving forwards and backwards. In DC power, electrons flow steadily in one direction. A hundred years ago, Nikolai Tesla and Thomas Edison fought it out over which type of electricity would be the basis for the grid, and Edison, backing AC, won. Both AC and DC can deliver energy over long distances – but at the time, AC was easier, so today we have an AC grid. However, your solar panels generate DC, your electric car motor is DC, and most of the machines in your home are probably DC. The only AC things likely to be left are an AC motor in your clothes dryer and perhaps another in the fridge, but even they will probably be DC soon if they aren't already.

We switch between AC and DC using machines (circuits, really) called "inverters", which turn the constant voltage of a DC machine into the wavy, oscillating voltage that is AC. If anybody tells you that "solar is killing the grid", they are probably referring at least in part to the need to "phase match" all the AC power that our solar inverters generate as it is fed back into the grid. Phase matching is just making sure all the AC waves line up. Think of all the inverters as singers in a choir. If they all match their tone and pitch, we get a lovely sound. It's the same with inverters. They need to match the voltage and phase of all that DC-generated power to the grid, otherwise the AC gets messed up and we start to have problems with machines that were designed to run on clean AC.

The grid won't be DC any time soon, so this need to convert DC to AC will continue. In assessing how much power your home can handle, one thing you will be asked by your sparkie is, "Do you have three-phase?"

Three-phase power: what is it and do you need it?

In your typical electrical cable, there are three wires. In a "single-phase" electrical system, there is an AC current on one of these three wires, while the other two are neutral and ground. Single phase is typically used for low-power systems and is the most common type of connection Australian households have to the grid. Three-phase systems, by contrast, send AC electricity in three phases, one on each wire. Three times as much power can therefore be sent over the same wires. Three-phase systems are more common in commercial buildings and light industry, where higher power is needed.

There is a limit to how much current wires can carry. As electricity passes through them, some of it is lost and becomes heat. If you put too much heat into the wire, it melts or shorts out. This is why wires are rated only up to certain currents. You can buy 10-amp, 15-amp and even 30-amp extension cords at the hardware store; the higher the current, the thicker the cord, because the copper inside them is thicker. If you try to put 30 amps through a 10-amp cord it will work for a short while, then it will heat up, and eventually melt and short out.

If you only remember one thing from high school science classes on electricity, it might be that the amount of power (P) you can get on a wire (measured in watts) is equal to the voltage (V) multiplied by the current (I) (measured in amps).

On a 230-volt system, the standard voltage in Australia, this means that on a 10-amp circuit, you can get power up to 2300 W, or 2.3 kW. A 30-amp circuit, which is about the heaviest grade you'll find in most

Australian homes, can support 6900 W, or 6.9 kW. The more amps you have, the more power your system can carry.

This is why your electrician might ask if you have three-phase power. A three-phase system triples how much power your circuits can handle. Let's look at a level 2 EV charger as an example. If you have a 30-amp, 230 V charger for your EV, on a single-phase system you'll get about 7 kW. Charging your car on such a system will give you 30–45 kilometres of range for every hour that you are plugged in, since most EVs get between 4 and 7 kilometres of range per kWh of charge.

On a three-phase, 230 V, 30-amp system, you can get 20,700 watts on the same charger, or about 21 kW. If you charge your vehicle for an hour, you'll get 100–150 kilometres of range.

For all but the most epic of road trips, level 2 charging is more than enough, even on a single-phase 7 kW system. Most people will not need to upgrade to three-phase if they don't have it already. But if you have a very large family, or a big house with a pool and lots of luxuries, or if you simply must run a 22 kW vehicle charger at home, you might consider upgrading your service to three-phase. To do so, you'll need to have a conversation with your local distribution company. There is often a charge, and you'll almost certainly need to upgrade your switchboard. You can expect this project to cost at least $2000–4000, but it could easily run past $10,000.

In terms of the bare minimum you need to know about electricity, I think it is this:

- You need to know that there is AC and DC. You can't mix them. It's like diesel and petrol. If you put petrol into your diesel car, the car is ruined. Similarly, AC will fry DC equipment and vice versa.

Fortunately, it is very hard to make this mistake, as their plugs and connectors are very different. Maybe you don't even need to know the difference between AC and DC – just that you shouldn't try to force one plug into another of the wrong shape. That's probably generally good life advice.

- You need to know that amps equate to power. The more amps, the more dangerous the machine. What kills people when they get electrocuted is the amps. Lots of volts at very low amps will merely scare the hell out of you.
- You need to know about watts (W). Watts are like horsepower. In fact, a kilowatt, which is 1000 W, is just a little bit bigger than a horsepower. The human body requires 100 W to walk. A human in peak condition can run at bursts of 400 W. An electric bike can comfortably move you around at 25 kilometres per hour using only 250 W. A toaster needs 1 kW, as does a kettle. A capable cooktop needs 3–5 kW to have as much power as its gas counterpart. You'll want 7 kW of power in a vehicle charger to charge quickly – or 22 kW to charge really quickly. A 14 kW solar system produces higher peak power than an 8 kW solar array. Over the course of the day, that extra power (kW) creates more energy (kWh).

FAQS

What about efficiency?

I'm going to say something that may sound heretical to many environmentalists and energy nerds: efficiency hasn't delivered, and electrification is now more important.

Efficiency has been in fashion since the oil crisis of the 1970s, which very few of us can even remember. America suffered worst because it was reliant for 15% of its energy on imports from the Middle East, which was now embargoed. Within days, the USA was in crisis. In the aftermath, the answer to the crisis that was most championed was efficiency. If all the (fossil-fuelled) cars and household boiler heaters were more efficient, the logic went, America could survive without Middle Eastern oil. And so efficiency became public policy, because conservatives and progressives could agree on one thing: waste is bad. This manifested as new efficiency standards for vehicles and home appliances. But there were a lot of problems and loopholes with these standards from the start. SUVs and pickup trucks weren't categorised as cars – they were "light commercial vehicles". While home appliances got more "efficient", with slightly better engines and gas burners, these gains were lost as appliances got bigger and people bought more of them.

The word "efficiency" was also applied to buildings, and we got very concerned about draughts and insulation. Absolutely, a well-insulated house uses less energy, is more comfortable and can even improve your health. We should mandate more efficient houses. The cheapest time to maximise efficiency is when you build a new house – the walls are open, and the underfloor will never be easier to access. But doing these things on a retrofit is very expensive, and for most people will not be the most effective way to spend money. The quote for retrofitting my 1930s bungalow with double-glazed windows was $120,000. It would take more than 200 years for that investment to pay itself back. Adding more solar and another heat pump is a cheaper near-term option, and I'm going to improve the insulation myself with cheap weather-stripping to eliminate draughts instead. That will do more, with a lot less.

If I thought about all the energy used in my home, and how much of it traditional efficiency could impact, I'd be sorely disappointed. These traditional efficiency measures really only apply to the heating and cooling components of my home, which are dominated by my water heater and space heaters. Because the water heater is already heavily insulated, more insulation would have only the tiniest marginal efficiency improvement. If I use heat pumps, which are more than twice as efficient as gas, for my space heating, I will lower my total energy use from around 100 kWh per day to around 90 kWh per day – around 10%. You might say that this calculation is unfair, because I'm including the energy used by my vehicle, which isn't affected by traditional insulation and efficiency measures, but I do that to emphasise that vehicles account for the majority of household energy use and emissions. I can improve insulation all I like, but if I don't also do something about my vehicle, my household emissions will remain high. If I electrify, including my vehicles, I'll see a huge drop in energy usage, to about 38 kWh. If I apply (lots of) insulation and double-glazed windows to my electrified 38 kWh, I might only get an additional 10% gain, moving my total household energy use to around 32–35 kWh per day.

The point is, electrification is the best tool we have to improve the energy efficiency of our lives. We should electrify everything as soon as we can if we are serious about efficiency, and we should apply strict performance standards to all our vehicles, including commercial vehicles. We should demand higher building standards, which will save the occupants far more money over the lifetime of the building than it costs to build the house more efficiently in the first place.

FAQS

What about recycling – what happens to old solar cells?

Many people worry about the resources used in electrification and whether they can be recycled. They often seem to do this after hearing some vitriolic misinformation on the internet.

First, let's put this question in the context of the waste we produce currently. On average, each Australian is responsible for around 15,000 kilograms of carbon emissions each year.* These are our "per capita" emissions, and they are among the highest in the world. Most of our emissions are invisible – although you might smell some of them in your kitchen or your driveway – but they all end up in the atmosphere. There is no recycling of CO_2 emissions.

All Australians also contribute in some way to municipal waste. This is your garbage, recycling and perhaps organic waste collection. It averages

---

* Hannah Ritchie and Max Roser, "Australia: CO_2 Country Profile", Our World in Data, https://ourworldindata.org/co2/country/australia

2200 kilograms per person each year.* Burning coal to make electricity creates a byproduct called "fly ash", which adds another 600 kilograms to every Australian's waste budget. Around half of our municipal waste and fly ash is recycled or recovered to make products such as concrete and metals. A tiny fraction of garbage is burnt to make energy. The rest is waste.

Compare this to renewables. Nearly all the materials used for electrification can be easily recycled. A solar module is mostly glass, aluminium and a little bit of silicon and silver. All these things are extremely recyclable. Nearly half (347 tonnes) of a wind turbine's weight is the steel in its tower. Most of the rest is accounted for by the gearbox and the motor (240 tonnes), which are made of steel, iron, copper and other recyclable metals. The hard to recycle bit of turbines are the (small percentage by weight of the machine) glass fibre composite blades. Investigators are researching wood, steel, aluminium and recyclable composites as alternatives. Probably 90% of the materials in solar cells and wind turbines are easily recycled. That makes them literally thousands of times better than fossil fuels per kWh of energy delivered.

At the household level, your car weighs 2 tonnes. Your water heater 180 kilograms. Your stovetop 20 kilograms. Your space heater 80 kilograms. Your rooftop solar might be 800 kilograms. In total, that's about 3 tonnes of material, almost all of it recyclable. Compared to throwing your carbon waste into the atmosphere, electrification makes for a much lighter footprint on the planet.

Renewables also produce vastly more energy for each kilogram of resources used. A 400 W solar panel weighs around 20 kilograms. A typical

---

* Department of Climate Change, Energy, the Environment and Water, https://www.dcceew.gov.au

5 MW wind turbine weighs around 800,000 kilograms.* The 400 W solar panel will likely last twenty-five years and create 14,000 kWh of electricity. The wind turbine will last twenty-five years and create 438,000 MWh of electricity. The solar panel makes 700 kWh per kilogram of materials used. The wind turbine makes 547 kWh per kilogram. By comparison, 1 kilogram of coal makes only 1 kWh of electricity. Renewables are many hundreds of times lower impact than fossil fuels.

If we choose to commit to electrification using materials that are inherently recyclable, we might become the best citizens of earth that have ever existed, recycling the products of our green energy system indefinitely. A colleague of mine was visiting BYD, a Chinese vehicle manufacturer; they described a credible plan to have a circular economy by 2040 – meaning 100% of the lithium, copper, steel and glass in their cars would be recovered and recycled.

And yes, given how much carbon your old fossil-fuelled machines emit, the environmental benefits of electrifying are immediate, not just long term. Even if the electricity that runs your electrified home still comes from a coal-generated power plant at first, every year, as we invest in more renewables, the emissions from the grid will go down. Eventually your electrified home will be running at zero emissions. Whereas if you buy an appliance or a vehicle that uses gas, petrol or diesel, it will always emit, for the rest of its life. Because the grid is decarbonising, every electrical asset gets cleaner over time.

---

* J. Jonkman et al., National Renewable Energy Laboratory, 2009, https://www.nrel.gov/docs/fy09osti/38060.pdf

FAQS

What about other climate actions?

Electrifying the machines in your life is the most effective action you can take to reduce our emissions and tackle climate change. But it's not the only one. I'm often asked about other actions and whether they can be similarly impactful. I'm wary of the "reduce, reuse, recycle" mantra of the 1970s because in order to deal with climate change, we need zero emissions, not fewer emissions. But obviously there are other changes we can make to our lifestyles to supplement the electric future and make it more sustainable globally.

What about my food?

Many people think that the best thing they can do for the climate crisis is to stop eating meat. It's true that what we eat has an impact on climate change. One factor is the amount of land required to grow animal feed. Another is that ruminants such as cows and sheep belch and fart methane, which is far worse as a greenhouse gas than carbon dioxide. Eating less meat remains one of the easiest consumer decisions you can make to reduce your climate impact – but it alone cannot solve our climate problem.

On an infrastructure scale, better land management and new low-carbon farming alternatives will lower the impact of occasional meat consumption. In Australia, we could try and substitute kangaroo meat, which has a very similar nutritional profile to beef but creates less than a tenth of the equivalent carbon emissions. At the moment, large numbers of kangaroos are culled and left to rot on farmland every year. Meat eating does not need to disappear completely, but Australians do need to become more conscious about their diets.

That said, the impact you can have on the climate crisis by electrifying your car and home is far greater than not eating meat.

What about my stuff?

Of course it's better for our planet if we all buy less stuff. I'm not so much talking about doing without, but about buying things mindfully, and buying good quality things that last.

I'm the last guy to talk to you about fashion (although I do have some pretty colourful shirts). But what I do know is that the fast-fashion industry convinces us that we need a constant churn of clothes. We all know that clothes and shoes stalk us online wherever we go. When we're bored, it's easy to push a button, auto-fill our credit card details and wait for a truck to bring the box to our house, only to wear the thing a few times before it falls apart or we get tired of it. Then, once you get rid of the piece of clothing, even if you donate it somewhere, most of it – 57% – ends up in landfill. Eventually, even if it's initially reworn, pretty much all of it will. And once the clothing hits landfill, it releases methane.

The fast-fashion industry is a big villain in the carbon-emissions picture. Current data shows that the industry emits approximately 1.2 billion tonnes of carbon dioxide annually. That represents around 10% of global greenhouse gas emissions, which is more than the combined emissions from the shipping and aviation industries.

A lot of fast fashion is made of polyester, which is now the most common fibre in the clothing industry. Approximately 70 million barrels of crude oil are used to make polyester, and polyester emits more carbon in its production than cotton. When you wash it, it releases microplastics into the wastewater, which end up in the ocean; about 25% of all microplastics come from synthetic fibres.

All this is not to say that you shouldn't buy new clothes if you like them – but be mindful and buy quality. In the USA, where fast fashion is a weekend sport, most people only wear about 10% of their clothes. The rest just take up space until they go to landfill. Instead, think about buying heritage pieces that you will wear forever, the way our grandparents did. Also, give new life to old clothes by buying from consignment and second-hand shops. Avoid synthetics if you can. People are doing amazing things with wool these days, which is breathable, doesn't stink, lasts a long time, and is sometimes compostable (check the label).

Much of what I've said about fast fashion also applies to many other products in our lives. A lot of furniture these days is built to furnish a house cheaply and isn't made to last. Instead of passing a sofa down to your children, you'll be lucky to get a few years out of the thing before it sags. I don't want to sound like an old curmudgeon (although I'm almost old enough to be one), but think carefully before you buy furniture, or a house, or anything that you bring into your home. Consider how long it will last. Invest in heirloom pieces. A solid wood table will last for generations and might be found at Vinnie's.

I'm not trying to be judgemental about buying stuff. I buy a lot of metal tools and toys. Just be mindful, and try to make sure that whatever you buy is stuff you will use, stuff you can repair if you need to, and ideally stuff that you can one day pass on.

FAQS

What about other energy sources?

Is renewable electricity our only option? There are other alternatives to fossil fuels, but for many reasons, electrification using renewable energy is by far the most effective and affordable way forward. I discuss the overwhelming arguments for renewable electricity in detail in my books *The Big Switch* and *Electrify*. If you would like to hear my breakdown of the pros and cons of renewables, nuclear, hydrogen, biofuels and more, you can listen to the audiobook recording of *The Big Switch* for free on the Rewiring Australia website. Chapter 4, "Australia's Energy Options", discusses the various alternatives.

From a scientific point of view, these debates have now been settled. If we allow ourselves to be distracted by them, we only delay what should be our top priority: electrifying everything and transitioning our homes and our electricity supply to clean, cheap, zero-emissions energy.

FAQS

What about farming?

There's a myth that electric vehicles are no good for rural life – but there's no reason why farms have to be run by diesel machines, and there are plenty of reasons for farmers to go electric.

My friend Mike Casey is the owner of the first all-electric farm in the industrialised world, if you don't count farms that never had electricity to begin with. He's also the CEO of NZO, or New Zealand Zed, a food-certification program that encourages farmers to go all-electric so that they can assure consumers that their produce was made without burning any fossil fuels.

Mike is an engineer and didn't set out to own a cherry orchard. But at a certain point in his life, he and his young family moved back to their homeland, New Zealand, and looked for a house close to Queenstown, where they could ski. They found a piece of land in Otago, and it was big enough that they figured they ought to plant something. They planted 9300 cherry trees.

Mike was already interested in electrification, and the first electric vehicle on their farm was a souped-up golf cart. Then he replaced the diesel pump used for irrigation with an electric one. He figures the payback period on that purchase is nine years. Then, he needed frost-fighting fans to keep the cherry blossoms from freezing. Those, too, he got electric, imported from South Africa. He then imported a small electric tractor from the USA to spray the foliage – the first electric tractor in New Zealand.

In total, Mike has twenty-one electric vehicles on his farm and saves a total of 60 tonnes of carbon emissions per year compared with a diesel-run farm. By installing solar and operating all-electrically, he has halved his energy costs, amounting to savings of around $100,000 per year. "The best way to be energy-resilient is to create your own energy," he says. He gets paid a 15% premium for his zero-carbon cherries. "And they taste like virtue," he says. Through his NZO certification system, other farmers can let consumers know that no carbon emissions were created when their produce was grown.

Most farms have many simple structures with large flat roofs, such as barns, that are perfect for generating cheap electricity. Yet until recently, farmers have been slow to adopt electric equipment. Why? "Quite simply, because we have been bashed to no end by the climate movement," says Mike. "Farmers are smart people who run multi-million-dollar, high-risk businesses. They are curious but conservative. Give them the economic reason, and they are there. Up until this point, there has not been an economic reason, until electrification became the clear winner."

Mike hopes that more farmers will follow suit and realise the savings involved in electrification, as well as the other benefits. "I have a happier, more engaged staff," he says. "The farm is safer, healthier, and there's less noise."

Farmers are already selling sunshine to lazy city-slickers . . . so why don't they electrify and sell them even more, at higher margins!

Conclusion: The Climate Why?

There is a lot of misinformation and disinformation – "miss and diss" – on the internet, and much of it casts doubt on climate change and its consequences. Within the scientific community, there is no doubt that climate change is real and a very, very serious threat. The only remaining disagreements now are about degree: how much, how fast and what to do about it. We know climate change is happening, we know it is costing lives and money and killing many species, and we know we should limit it as much as we can.

As a result of climate change, sea levels are rising, the earth's average temperature is rising, ecosystems are changing, and there are more frequent and more costly natural disasters. The earth has already warmed by 1 degree and is approaching a critical target of 1.5 degrees,* a figure we

---

* In fact, we have already surpassed this average for one year, although we haven't yet surpassed it permanently. See Mark Poynting, "World's First Year-Long Breach of Key 1.5C Warming Limit", BBC News, 8 February 2024.

will unwisely cross sometime this decade or early next. Over hundreds of years, 1.5 degrees of warming will raise the sea level by 2 to 3 metres. Warming of 2 degrees will raise sea levels by 2 to 6 metres. Five degrees of warming will cause sea levels to rise by 20 metres. These projections are long-term. In the nearer term, seas are projected to rise by between 30 centimetres and 1 metre by the end of this century on the NSW North Coast.

Around 80% of our emissions come from our energy sources, with the other 20% coming from agriculture (including livestock), deforestation, other changes in land use, industrial chemicals and other sources. The reason people talk about an "energy transition" is because we need to eliminate these emissions by 2040 or 2050 at the latest, to have any chance of keeping climate change to between 1.5 and 2 degrees of warming.

Where can we get enough energy to replace our fossil fuels? As we have seen, our only real options are renewable energy sources, supplemented perhaps by nuclear energy in some countries where renewables aren't readily available. Because our zero-emissions energy sources are mostly electric, electrification is the main tool in our climate-fixing toolbox. Electrification is our only viable option at a planetary scale, and it is incredibly efficient. If we electrify everything, we will only need half (or less) of the energy we need today to do all the things we currently do, only cleanly.

You have probably heard someone, perhaps a talkback radio host or one of your uncles, say that Australian emissions don't count, or that the emissions from our cars don't count. They're wrong: all carbon emissions contribute to the climate crisis. But people love to manipulate statistics, and sometimes the data behind those statistics needs to be interrogated.

The methodology of the IPCC (Intergovernmental Panel on Climate Change) is to count emissions in the country where they occurred, not in the country of origin of the fuel that created them. Australia's emissions in 2019 counted as 554 Mt (millions of tonnes). Since we produce and export so much coal and other fossil fuels, we are also the source of an additional 1289 Mt of emissions. But because those fuels are burnt overseas, they don't count as ours. Of the 554 Mt of emissions that are counted as Australia's, I estimate that 224 Mt, or 40%, support our exports – for example, the diesel used in Australian mining to get our ores to overseas markets, and the 60% of our beef that is grown here and produces methane here before being exported and eaten overseas. When it comes to carbon emissions, we are truly globalised, and the effects of those emissions are also global.

Our fossil fuel exports mean that our small population is an outsized contributor to global climate change. Consecutive federal governments haven't stopped the exports, despite the glaringly obvious climate science. Because the IPCC methodology doesn't count those export emissions against us, governments can pretend it isn't their problem – except it is, because it is our collective carbon problem.

On top of this, Australians aren't getting paid what those resources are worth. Most of the companies extracting our resources are owned overseas, and the profits go overseas, typically without the companies paying any tax.* And because we didn't design the resources rent tax very well (or, rather, it was designed by fossil fuel lobbyists to benefit their interests), we earn about $100 million on a volume of gas that would earn

---

* Ian Verrender, "How Your Spiking Energy Bills Are Making Foreign Investors Rich", ABC News, 13 June 2022.

Qatar $32 billion. We are being stiffed.

A relatively tiny number of FIFO workers get a great pay cheque from the fossil fuel industries, but there are far more jobs to be had in creating the technologies and energy sources of the future. It's long past time for Australia to give up this deceitful charade in partnership with the international fossil fuel industry. We have enough fossil projects to deliver the fuels we'll need to get us to 2040, by which time we should be largely electric and zero-emissions. There is no need to approve new fossil fuel projects – and governments should be toppled if they do.

Of the 331 Mt of emissions produced to run the domestic economy – the things we do for Australians in Australia – the largest fraction, 140 Mt, is household energy use, including our electricity, vehicles and gas. These household uses account for around 42% of our domestic emissions. The commercial sector – our small businesses, office buildings and commercial vehicles – are another 96 Mt or 29%.

For the most part, decisions about how we use energy in our homes and small businesses are made around the kitchen table. So somewhere between 42% and 71% of our emissions – let's call it half – are determined by our household purchasing decisions. The rest are the result of decisions made by governments and corporations – beyond our direct control, but still potentially influenced by our decisions about what to buy and how to vote.

So, why rewire your life? Why rewire the world? Every Australian emits over 15 tonnes of carbon dioxide per year. Americans emit 16–17 tonnes, Japanese and Germans just under 10 tonnes, Chinese about 7.5 tonnes, and Indians about 2 tonnes. We know that these carbon dioxide emissions are heating the earth because they trap more of the sun's incoming sunlight. Scientists have known that they would do this for more than

100 years. It has been understood by American presidents since 1965.* And it has been in the public consciousness since at least 1988, when the renowned climate scientist James Hansen testified to the US Senate on the issue.**

We all have a role in this transition, beginning with our households. If we make those decisions wisely, we will also lower our living expenses, improve our health outcomes and compel industry to solve the other half of our emissions dilemma.

Sounds good, doesn't it? Let's get on with it.

---

* The White House, *Restoring the Quality of Our Environment: Report of the Environmental Pollution Panel President's Science Advisory Committee*, Washington DC, 5 November 1965.

** Philip Shabecoff, "Global Warming Has Begun, Expert Tells Senate", *The New York Times*, 24 June 1988.

Acknowledgements

Laura Fraser provided narrative solutions. Josh Ellison told stories with data.

Thanks to the model electrifiers: Bryn Foletta, Katherine McConnell, Karl Jensen, Sarah Aubrey, Dan Bleakley, Dan Adams, Kim Loo, Fred Hopley, Chris Kerr, and Thor and Hannah Denmark.

Thanks to serious book people Chris Feik and Denise O'Dea.

Thanks for being generous and helpful: Pamela Griffith, Ross Griffith, Dan Cass, Francis Vierboom, Kristen McDonald, the Electrify 2515 team, the Rewiring Australia team, the Rewiring America team, Sam Calisch, Vince Romanin, Ty Christopher, John Buchelin and Mike Casey.

And to my wife: I couldn't have written this without Arwen, who shares my passion for making the world better for everyone and is far less fussy than I have made her out to be for narrative purposes.

Resources

General electrification advice

Rewiring Australia
rewiringaustralia.org
A non-profit, non-partisan organisation dedicated to representing the people, households and communities in the energy system, founded by Saul Griffith and Dan Cass. Visit our website for leading data, research and advocacy on electrification.

Your Home
yourhome.gov.au
Advice from the federal government on how to make your home and appliances more sustainable.

Home Scorecard
homescorecard.gov.au
Calculate your home's current energy use and receive tailored suggestions for improvements.

Switched On

switchedon.reneweconomy.com.au

A website and podcast focusing on household electrification, including "news and analysis, myth busters, podcasts and web stories that identify the opportunities and challenges of electrifying everything".

My Efficient Electric Home Facebook group

A great place to ask questions and research how others have gone about their electrification journey.

Climate Council

climatecouncil.org.au

Resources including a home appliance savings calculator from an independent, evidence-led Australian climate organisation.

Renew

renew.org.au

A non-profit focused on "transforming Australian homes for climate and energy resilience since 1980", Renew runs events and offers advice about sustainable living, including solar, batteries and how to get off gas.

One Million Women

1millionwomen.com.au

A home electrification guide from "a global movement of women empowering them to act on climate change through the way they live".

Energy Consumers Australia

energyconsumersaustralia.com.au

An independent group representing residential and small business energy users. Their consumer resources include a guide to going "all-electric".

Solar Citizens
solarcitizens.org.au
Resources and advocacy for solar-powered households.

Home Assistant
homeassistant.io
A website devoted to "open-source home automation that puts local control and privacy first. Powered by a worldwide community of tinkerers and DIY enthusiasts".

Solar and batteries

The Energy Saver Handbook
savewithsolar.org.au
A free guide to reducing your energy use, cutting your energy bills and making the most of rooftop solar.

SunSPOT
sunspot.org.au
Built by engineers and data analysts from the University of NSW, SunSPOT helps you calculate what kind of solar system your household needs and how much it might save you.

Solar Quotes
solarquotes.com.au
An independent platform where you can compare solar quotes from installers across the country.

Solar Choice
solarchoice.net.au
Another solar quotes comparison platform, Solar Choice also provides buying guides for electric vehicles, heat pumps and batteries.

Wattblock
wattblock.com
Wattblock offers advice, resources and case studies for electrifying strata properties, including batteries, EV charging and hot water.

Getting off gas

The Getting Off Gas Toolkit
gettingoffgastoolkit.com
Practical advice prepared by Renew.

Getting Off Gas: Your Four Step Plan
brighte.com.au/getting-off-gas
A step-by-step guide to switching from gas to electric.

Electric vehicles

Electric Vehicle Council
electricvehiclecouncil.com.au
A national body representing the EV industry, the EVC publishes consumer guides to buying and running your EV.

The Australian Electric Vehicle Association
aeva.asn.au
A non-profit organisation representing "EV consumers, users and enthusiasts", the AEVA publishes resources for EV drivers and runs events, with branches in every state and territory.

Cooking

How to Buy a Great Induction Cooktop
https://www.choice.com.au/home-and-living/kitchen/cooktops/buying-guides/induction-cooktops
An induction cooktop buying guide by CHOICE, "the leading consumer advocacy group in Australia".

Renters

A Renter's Guide to Sustainable Living
renew.org.au/publications/renters-guide-to-sustainable-living
A handbook for renters prepared by Renew.

Make the Switch: Rental Homes

maketheswitch.org.au

Advice for tenants and landlords on how and why to electrify rental properties.

Better Renting

betterrenting.org.au

A community organisation that provides resources, advice and advocacy for more sustainable and affordable rental homes.